INDIE: A Field Guide to AI-Augmented Learning & Development

INDIE

INvestigation-Design · Development · Implementation · Enhancement

A Field Guide to ***AI-Augmented*** *Learning & Development*

David Q. Dauthier

Senior Instructional Designer

Poikilos Press

NOSTOS, LLC

INDIE: A Field Guide to AI-Augmented Learning & Development

Cover by Claude Design, Q & KWD.
Illustrations by Grok & Q.
Typeset in Palatino Linotype

ISBN: 979-8-9870802-0-7

First Edition, 2026

Printed in the United States of America

This book is dedicated to

Ellen, Mary, and Donna,

Three Generations of Mothers

Who, amidst Trials and

Tribulations, made me.

"In our world," said Eustice, "a star is a huge ball of flaming gas."
"Even in your world, my son," said Ramandu, "that is
not what a star is but only what it is made of."

The Voyage of the Dawn Treader

CS Lewis (1952)

"AI is just autocomplete."

Overheard at a Christmas Party

Anonymous (2025)

Contents

Foreword

I did not begin my professional life as an instructional designer. I began it as a teacher of classical languages. From 1998 to 2018, I taught in China. In those two decades, among other things, I taught Classical Latin and Ancient Greek at the Graduate School of The Chinese Academy of Social Sciences (中国社会科学院研究生院). I also worked at The High School Affiliated to Beijing Normal University (北京师范大学附属中学), where I taught English Composition, Literature, and Latin. I continue to serve as an advisor to *Latinitas Sinica,* a national-level institute for the promotion of Latin and Ancient Greek studies in China, operating under the auspices of Beijing Foreign Studies University (北京外国语大学).

Over the course of those twenty years, I stood in front of thousands of students whose native language was, for the most part, Mandarin and whose educational traditions were formed in the context of an ancient civilization. I asked them to master the languages, and to be knowledgeable of the cultures and histories, of Rome, Athens, and the Anglosphere. This was an education not only for them, but also for me.

One of the many things I discovered in my tenure was that textbooks and curricula are often designed by people who have never taught, or who taught without ever grasping the difference between knowing a subject and knowing how to help someone learn it. My own training was in languages, not education, so I read widely to close that gap, first out of professional necessity and later out of genuine curiosity. I moved into instructional design as a profession in 2020, after completing my graduate studies in the field, bringing a set of tendencies that function as a kind of professional conscience. On one side sits a deep suspicion of educational content that is not grounded in learning theory, and on the other, a respect for how genuinely hard it is to change a person's behavior. These tendencies cause me to insistent that training do what it claims to do and that it be measured honestly against that claim, so when AI was introduced to me as a tool for helping achieve those ends, I took it seriously and set out to understand it at depth, from first principles. To my surprise, I found that I happened to have a significant head start.

The Mathematics of Emergence

Bear with me if math isn't your thing. It's a hobby of mine, and it taught me something about AI that I want to share, but it needs a little background first.

In the late 1990s, I became fascinated with what could be called a mathematical backwater: two-dimensional cellular automata. The idea is simple. Picture a grid of cells, like a chessboard, where each cell can be in one of two states, on or off, like an old-fashioned lightbulb. The cells blink on and off according to a few simple rules that depend only on the states of their neighbors. *Conway's Game of Life*, the most famous example, runs on such a binary grid under just four elementary rules.

What those four rules produce is anything but elementary: self-replicating structures, patterns that fly across the grid, oscillators, and emergent behaviors so rich and varied that mathematicians and computer scientists have spent decades cataloguing them. The rules are simple; the universe they create is not, and the complexity of that universe cannot be fully derived from the simplicity of the rules that produce it. I realized that. AI is much the same. Those who insist that it is just autocomplete are right, of course, but they miss the point. If you ask me who I am, I won't begin by telling you that, among other things, I'm 170 pounds of elemental oxygen. It's true, it is what I'm made of, but it's no more what I am than autocomplete is what AI is.

In 1960, the physicist and mathematician Eugene Wigner published an essay called *The Unreasonable Effectiveness of Mathematics in the Natural Sciences*. In it, he observed that mathematical structures born in purely abstract contexts, with no practical application in mind, have an uncanny tendency to describe physical reality with extraordinary precision. He found this deeply mysterious, and he was right to. His essay has lost none of its strangeness in the decades since.

When I set out to understand artificial intelligence at a foundational level, I worked through a textbook on the mathematics of machine learning. What it showed me was not what I expected. I expected demystification. What I found was deeper mystery. The effectiveness of large language models, the way statistical pattern recognition across unimaginable quantities of text yields something that reasons, synthesizes, and generates language of startling quality, is, in Wigner's sense, unreasonably effective, and no one is actually sure why it works at all, much less so well.

My background in mathematics gave me the ability to recognize something. AI is not a passing fad. This is not a more sophisticated search engine or a better autocomplete. It is a qualitative change in the nature of tools, of the kind that happens rarely and matters enormously. I had seen simple rules produce complex universes before, and I recognized the pattern.

Serious Tools for Serious Work

I work as a Senior Instructional Designer in the utilities sector. The people I help train work with natural gas and electricity. Used properly, they fuel the engine of modern civilization. Used improperly, they can devastate whole communities. In much of the work we do, there is no margin for error. There is no version of this work where a person fakes competence until they develop it. Someone who is faking can get themselves or others killed. I take my work very seriously.

This shapes how I think about training. Instructional design is fundamentally about behavior modification. The evaluations don't matter, and the completion rates don't matter if what we build does not change what people actually do. The appearance of learning is not learning. In a high-stakes regulated environment, the difference between the two is visible in ways that more forgiving contexts conceal. If you design learning well, you know it works. If you design it poorly, you know that, too.

When AI tools arrived and began dripping into our workflows, we treated them the way we treat any new tool: seriously, carefully, and with an immediate question about standards. It quickly became apparent that this tool was unlike any we had been given before. Its capabilities were of a different order. Its failure modes were of a different kind. If we were going to use AI correctly, there would have to be guidelines, systems, and a framework for surfacing best practices and improving them continuously. I set myself to that task. The book you are holding is the result.

Why This Book

The specific moment the book became necessary was not a dramatic one. There was no single conversation, no sudden realization. It came about by watching colleagues, people who understood learning theory and could build programs that worked, engaging with AI tools without any framework adequate to what those tools actually were. The responses I observed clustered at two poles: uncritical enthusiasm that treated every AI output as a gift, and reflexive anxiety that treated every AI capability as a threat. This caused me concern.

That, more than anything else, is what crystallized the need for the book, not the tools themselves, but the professional response to them. The field of instructional design had produced, over fifty years, a body of rigorous thinking about how people learn and how to design experiences that support learning. That thinking process is not invalidated by AI.

On the contrary, it was made more necessary by it. What was missing was a framework that connected AI-augmented practice to that theoretical foundation, a framework that was honest about what the tools changed and what they did not, a framework that was specific enough to guide decisions on an actual project that had integrated AI tools.

I named this framework INDIE, for **IN**vestigation-Design, **D**evelopment, **I**mplementation, **E**nhancement. The name is also, I admit, a small act of professional conviction. This is a framework for practitioners who think for themselves.

To the Practitioner Reading This

You may be excited as you read this, or you may be anxious. Most likely, you are somewhere in the middle, aware that something significant is happening and uncertain what it means for the work you have spent years building the capacity to do. I want to say something to you directly, before the frameworks and the prompts and the examples come rolling in. It's old advice:

Noli timere.
Do not be afraid.

This book offers no reassurance that the anxiety you feel is irrational; some of it is clearly rational. Chapter 3 will address the anxiety honestly, separate what is legitimate from what is distortion, and will not brush aside the real concerns.

The arrival of AI tools means, among other things, that the time of the pure production specialist has passed, and the era of the strategic practitioner has arrived. The tasks that none of us love: the busy work, the first-draft production, the formatting, the synthesis of thirty survey responses into a coherent theme map, much of that can now be delegated. You are no longer, in the old sense, at the bottom of the production chain. You now have an assistant that is, within its domain, a genius. It does not have your judgment. It does not know your organization, your learners, your subject matter, or the political dynamics of the stakeholder conversation you are about to walk into. But it can draft, synthesize, generate, and analyze at a speed that frees you to do the work that actually required *you* all along.

Getting the most from that assistant requires understanding it. Not at the level of software engineering, but at the level of professional practice: knowing what it is good for, knowing where it fails in ways that matter, knowing how to give it the context that makes its outputs genuinely useful rather than generically plausible. That is what this book is for.

A Word on Standards

One thing does not change, and I want to say it plainly: The instructional designer who uses AI to produce training that is not rigorously designed, that has not been evaluated against the theoretical standards the field has developed over better than half a century, that has not been reviewed by someone with genuine domain expertise, that cuts corners because AI makes cutting corners fast and invisible, bears full professional responsibility for what that training produces in the people who encounter it.

While AI can accelerate production, synthesize data, draft content faster than any human, it cannot be held responsible for the quality of what it produces. You can. Times change. Standards must remain. While it's a bit harsh, I remind myself and my teammates that "AI cannot be fired for its mistakes, but you can."

This book is written for the practitioners who take this work seriously. The ones who know that a learning program is not finished when it is delivered; it is finished when the people who experienced it can do something they could not do before. The ones who will use AI to become more of what they already are, rather than less. The ones who understand that a powerful tool in careful hands is an instrument of genuine human development.

This book is for you.

Acknowledgments

Before we go an inch further, this book itself is proof positive of all the claims that will be put forth in it. This book was designed and developed with the help of artificial intelligence. That's not a confession; it's a demonstration. Right now, I would like to thank Grok, Claude, and Gemini for sparring with me, acting as drafting partners, editors, and critical reviewers throughout this project. They pushed back; they suggested; they drafted; and they were wrong in ways that required me to think more carefully than I would have on my own. Every limitation this book identifies in AI tools was encountered firsthand in the writing of it. A major claim of this book is that the human element is essential, so it is to the human element that I now turn.

Several of my colleagues not only reviewed various parts and versions of this manuscript but helped test the INDIE process in real projects. Their willingness to engage with this work seriously, to bring their professional judgment to bear on something that was still being formed, and to tell me honestly when it wasn't working is present on every page. I am indebted to Mr. Brad Aungst, who in addition to excellent feedback has been a source of ideas and continual encouragement, Mr. Kyle Robie, Ms. Joanna Wroblewska, Mr. Garrik Hudson-Falcon, and Mr. Michael Robinson. I would be remiss if I did not mention Mr. Troy Nutter, who is indirectly responsible for much of this work by virtue of his forward thinking, high standards, and direction in all matters L&D.

A very special thanks to my oldest friend, Mr. Kurtis D. Hamm, who went through the manuscript line-by-line, providing comment at every stage. I would like especially to thank Mr. Kevin Woo for his advice, support, and absolutely everything else. Last, but certainly not least, a very special thanks to Miss Fufu for both being there and being tolerant.

All mistakes that remain in this book are mine alone.

David Quentin Dauthier

Bothell · Washington · 2026

Introduction

Something significant is happening in learning and development, and it is happening faster than the field's frameworks have been able to absorb. In the span of a few years, artificial intelligence has moved from a speculative presence on conference keynote slides to an active participant in the daily work of instructional designers, learning experience designers, facilitators, performance consultants, and L&D managers around the world. AI tools are drafting content, synthesizing research, generating assessment items, producing learning objectives, analyzing feedback data, and supporting the design of programs at a speed and scale that would have seemed implausible even at the start of this decade.

The field's response has been, understandably, uneven. Some practitioners have embraced the tools with enthusiasm that has occasionally outrun their critical judgment. Others have held back, waiting for clarity that has been slow to arrive. Most have found themselves somewhere in the middle: using AI tools with varying degrees of confidence, uncertain about which practices represent genuine progress and which represent shortcuts that will show up as quality problems later.

What has been largely missing from this moment is not more enthusiasm or more caution. It is a rigorous, practical framework: something grounded in learning theory and professional standards, honest about what AI genuinely changes and what it does not, and specific enough to guide decisions in real projects under real constraints. This book is that framework.

What This Book Provides

INDIE: Investigation-Design, Development, Implementation, Enhancement, is a complete instructional design framework built for the conditions that exist now, not the conditions that existed when ADDIE was codified in the 1970s. It does not reject what ADDIE got right. It updates the architecture for a world where iteration is fast, content production is cheap, and the pressing professional challenge is no longer how to produce training efficiently but how to ensure that what is produced is genuinely effective.

The framework integrates AI at every stage, not as an add-on to existing practice but as a structural element of how modern learning design works. It addresses the tools practitioners are actually using, the quality risks those tools introduce, the governance and ethics obligations that come with them, and the organizational change work required to shift a team or a function from ADDIE-based practice to something better suited to the current environment.

At the same time, the book is grounded in what the field has always known about how people learn: the theoretical work of Gagné, Knowles, Vygotsky, Mager, Bloom, Kirkpatrick, and Gilbert. These frameworks do not become obsolete when AI enters the workflow. They become more important, because the speed at which AI produces content makes the quality of the pedagogical judgment applied to that content more consequential, not less. A poorly designed learning objective used to be an expensive mistake. In an AI-augmented workflow, it becomes the seed of an entire poorly designed program, produced at high speed.

> *The question is no longer whether AI will be part of instructional design practice. It is whether practitioners will lead that integration or follow it.*

For Whom Is This Book Written?

The book was written with instructional designers as the primary audience. But the work of learning and development is done by a much wider range of practitioners, and every role in that ecosystem is addressed directly.

Role	What this book provides
Instructional Designer	A complete framework for AI-augmented practice: how to structure the work, which tools to use at which stages, how to maintain quality standards, and how to develop the professional judgment that AI cannot replace.

Learning Experience Designer	An account of how the LXD orientation of deep understanding of learners' job realities and performance contexts, emotional arc, and experience quality integrates with AI-assisted production workflows without being displaced by them.

eLearning Developer and Multimedia Producer	Specific guidance on AI tool integration into production workflows, quality standards for AI-generated media, and the review practices that prevent the failure modes unique to AI-assisted production.

Facilitator and Trainer	A practical account of where AI genuinely serves facilitation preparation and post-session synthesis, and where the irreducibly human act of being present in a room cannot be accelerated.

Performance Consultant and OD Specialist	AI applications for diagnostic work, hypothesis testing, data synthesis, and organizational change, alongside an honest account of the analytical courage no tool can supply.

L&D Manager and Curriculum Coordinator	Portfolio management, team capability development, AI governance frameworks, and the organizational change work of transitioning a team from ADDIE to INDIE, with a complete 18-month adoption roadmap.

Learning Technologist	A repositioning of the learning technologist role from post-development support to IN-phase collaborator and Enhancement-phase data custodian, the infrastructure role on which AI-augmented quality depends.

Solo Practitioner and Consultant	An honest account of what AI has changed for independent practitioners: the capacity gaps it closes, the quality risks it introduces, and how to build a sustainable AI-augmented practice without a team.

CLO and Senior L&D Leadership	The strategic framing for AI integration in learning functions: what changes about program economics, what the governance obligations are, how to make the organizational case for the investment, and what an AI-native L&D team looks like.

The book also speaks to the adjacent practitioners who make L&D work: subject matter experts who review training content, project managers who hold project timelines together, HR business partners who commission learning programs, and organizational leaders who sponsor development initiatives. A chapter addresses each of these roles directly. Every practitioner who has ever found themselves in a project room with an instructional designer will find something here that clarifies what the collaboration requires and what it can produce.

Why This Book, Why Now

There is no shortage of writing about AI in learning and development. Conference sessions proliferate, vendor white papers accumulate, LinkedIn feeds fill with takes that range from the utopian to the alarmed. What has been scarcer is writing that is simultaneously rigorous and practical: grounded in evidence, honest about uncertainty, specific enough to be used on an actual project on an actual Monday morning, and substantial enough to remain useful as the tools continue to evolve. This book was written to fill that gap. Several features distinguish it from the existing literature:

- **It is grounded in learning theory.** Every framework and recommendation in the book is connected to the theoretical work that gives it its warrant. Practitioners who understand why a design decision is right, not just that it is recommended, make better decisions when the specific situation departs from the template.
- **It is honest about what AI cannot do.** The book does not argue that AI will transform instructional design into something unrecognizable. It argues the opposite: that AI makes the specifically human dimensions of this work more important, because the volume and speed of AI-assisted production makes the quality of human judgment applied to that production more consequential.
- **It addresses governance and ethics directly.** A dedicated chapter on AI governance, bias, data privacy, intellectual property, and the practitioner's ethical responsibility is not an afterthought. It is

the framework's moral foundation, placed where it belongs: before the book ends, not in an appendix.

- **It includes a complete prompt library.** Appendix A contains 34 prompts organized by INDIE phase and task, each with a rationale and a statement of known limitations. These are not generic AI prompts. They are instructional design prompts, engineered for the specific analytical and production tasks the framework requires.
- **It will remain useful as the tools evolve.** The specific AI tools available today will change. The framework they are used within will not. INDIE's structure, the theoretical grounding, the quality standards, the governance principles, and the professional judgment the book develops will apply to the next generation of AI tools as they apply to this one. This is a book for the next several years of practice, not the next several months.

The Field This Book Is Entering

Learning and development is, by any reasonable assessment, at an inflection point. The combination of AI-accelerated content production, organizational pressure for faster and more demonstrably effective learning, and the arrival of adaptive delivery systems that can personalize learning at scale is changing the economics of the field in ways that will reshape it substantially over the coming decade.

The practitioners who will lead that reshaping are not necessarily the ones who adopt AI tools most quickly. They are the ones who understand what the tools are for, what they require in terms of human oversight and judgment, and how to build the organizational conditions that allow AI to accelerate quality rather than simply accelerate output.

That understanding is what this book is designed to develop. It is written for practitioners who take their work seriously, who know that the measure of a learning program is not whether it was delivered efficiently but whether it produced genuine competence in the people who experienced it, and who recognize that AI is the most significant change in the conditions of that work since the field was founded.

The field has navigated significant transitions before: from classroom to eLearning, from eLearning to mobile, from periodic training to continuous learning ecosystems. It has navigated them by returning, each time, to the foundational questions: what does the learner need? What does good instruction require? How do we know whether it is working? Those questions do not change. The tools for answering them do. This book is a guide to answering them with the tools that now exist, at the standard the field has always demanded.

How to Use This Book

The book is organized in five parts and eighteen chapters, followed by five appendices. It is designed to be read straight through, but it is also designed to be entered at any point depending on your current need.

- **If you are new to AI tools in L&D practice,** begin with Part One (Chapters 1 through 4). These chapters establish what AI is, what the professional anxiety surrounding it reflects, and how to work with AI tools effectively before the framework itself is introduced.
- **If you want the framework quickly,** go to Part Two (Chapters 5 through 9). Chapter 5 presents the case for INDIE over ADDIE. Chapters 6 through 9 address each phase in full detail, with prompts, examples, and quality standards for each.
- **If you are looking for guidance specific to your role,** go to Part Three (Chapters 10 through 16). Each chapter addresses a specific L&D role: its unique relationship to AI, its specific AI applications, and the professional judgment it must supply that AI cannot.
- **If you are a leader planning an organizational transition,** Chapter 18 presents a complete transition roadmap with the six most common failure patterns and how to prevent them. Chapter 17 addresses the governance and ethics obligations that the transition requires.
- **If you need a prompt immediately,** go to Appendix A. The prompt index at the front of the appendix will direct you to the

right prompt for your current task. Read the "known limitations" annotation before using any prompt in a high-stakes context.

A Note on Audience and Tone

This book takes its readers seriously. It does not simplify what is genuinely complex, and it does not offer reassurance that is not warranted. Where AI tools introduce quality risks, those risks are named specifically. Where the professional challenges of an AI-augmented field are real, they are addressed directly rather than dissolved in optimism.

It also does not speak to the lowest common denominator of L&D practice. It is written for practitioners who are already competent, who are already committed to quality, and who need a framework that matches the sophistication of their professional ambition. If you are looking for a book that will make AI in L&D seem easy, this is probably not it. If you are looking for one that will make it tractable, rigorous, and genuinely useful, read on.

The Argument in Brief

This book makes one central argument, developed across eighteen chapters and supported at every step by practical frameworks, worked examples, and specific prompts:

> *AI changes the economics of instructional design practice. It does not change what good instructional design requires.*

What it requires has always been clear: understand the learner, identify the right gap, design the experience that closes it, implement it in conditions that allow it to work, and measure honestly whether it did. AI accelerates several of these steps and supports others. It substitutes for none of them. The practitioner who understands this clearly is the practitioner who will use AI well, benefit from what it genuinely offers, and hold the line at what it cannot replace.

That practitioner is this book's reason for existing.

INDIE

INvestigation-Design · Development · Implementation · Enhancement

A Field Guide to AI-Augmented Learning & Development

Part One · Chapter 1

The Profession at the Inflection Point

Maria sat at her desk on a Tuesday morning with a cup of coffee that had gone cold, looking at a learning objective she had just generated in 30 seconds.

She had been an instructional designer for eleven years. She had written thousands of learning objectives. She knew what a good one looked like. She knew what a bad one looked like. She knew the particular satisfaction of writing one that was genuinely precise, that connected to a real performance gap, that could be assessed in a way that meant something. Writing a good learning objective had taken, on average, about twenty minutes.

The one on her screen was good, very good. She had checked it against Mager's criteria. It was specific, it was measurable, it connected to the job task she had analyzed the previous afternoon. And it had taken 30 seconds. She sat with that for a moment.

The question forming in her mind was not "is this cheating?" She had moved past that question, because she recognized it as the wrong question. The right question was harder to articulate, and it was not going away: if I can do this in 30 seconds, what exactly is my job now?

The Profession that Arrived Here

Instructional design is a young profession by historical standards. Its theoretical foundations were laid in the mid-twentieth century, its professional infrastructure consolidated in the latter decades, and its current tools and frameworks are still, in many respects, products of decisions made when the central technological challenge was the transition from classroom instruction to slide projectors and film strips.

This is not a criticism. The field produced rigorous thinking about how people learn, how instruction should be structured, how objectives should be written, how assessment should connect to learning, and how programs should be evaluated. The theorists whose work underpins contemporary instructional design, Gagné, Mager, Bloom, Kirkpatrick, Knowles, did work that remains valid. The questions they asked are still the right questions.

What the field has carried forward alongside that rigorous thinking is a process model, ADDIE, that was designed for a specific set of economic and technological conditions. Those conditions have changed. The process model has not kept pace.

Period	Development	What Changed
1940s–1950s	**Systematic training design emerges**	The U.S. military, facing the need to train millions of people rapidly, develops systematic approaches to instructional analysis and sequencing. Robert Gagné and others lay the theoretical groundwork that will define the field.
1960s	**Performance objectives enter the field**	Robert Mager publishes "Preparing Instructional Objectives." For the first time, the field has a shared language for stating what learning is supposed to produce. The measurable objective becomes the cornerstone of instructional design practice.
1970s	**ADDIE codified**	Florida State University develops what becomes known as ADDIE for military training contracts. Analyze, Design, Develop, Implement, Evaluate. A logical, defensible sequence that matches how people intuitively think about building things.
1980s–1990s	**ADDIE spreads**	As instructional design professionalization grows, ADDIE spreads across corporate training, higher education, and government. Its sequential structure maps neatly to project management frameworks. It becomes the default answer to "how do you do instructional design?"
2000s	**eLearning arrives**	Digital delivery changes production workflows but not the underlying ADDIE sequence. The analysis deliverable still precedes the design deliverable. The design deliverable still precedes development. Evaluation still happens at the end.
2010s	**Agile and rapid development**	The field begins questioning ADDIE's linear assumptions. Agile instructional design frameworks emerge. Practitioners note that the separation of analysis from design, and design from development, creates costly late-stage rework. The critique is correct. The alternative frameworks are incomplete.

2020s	**AI changes the production equation**	Generative AI collapses the cost and time of content production. The economic logic that justified ADDIE's sequential structure no longer applies. The field needs a framework built for the conditions that now exist, not the conditions that existed in 1975.

What ADDIE Got Right, and Where Its Logic Breaks

It is worth being precise about this before introducing INDIE, because the argument is not that ADDIE is wrong. The argument is that ADDIE's logic was sound for the conditions it was designed for, and those conditions have changed in ways that make its sequential structure counterproductive.

ADDIE's logic was economic. When development was expensive, you did not begin developing until the design was complete. When the design was expensive to change, you did not begin designing until the analysis was complete. When evaluation happened at the end, the program had to be substantially built before you could learn whether the design was working. Each sequential handoff made sense as a form of quality control in an environment where iteration was very costly.

AI collapses the production cost. A prototype that previously took two weeks of developer time can now be produced in hours. A scenario set that previously required multiple rounds of content development can be generated and evaluated in a single session. The economic argument for sequential phases disappears when the cost of building and rebuilding falls by an order of magnitude or more. You can afford to design and learn simultaneously when iteration is cheap.

The second structural problem is the fiction of sequential completion. In practice, analysis and design have never been truly sequential. Designers make design decisions during analysis, and analysis discoveries force design revisions. ADDIE described an ideal sequence that practitioners regularly violated because the work itself required it. The violation was not bad practice; it was an acknowledgment that learning design is iterative by nature. INDIE makes the iteration explicit rather than treating it as a deviation from the model.

The economic logic that made sequential phases sensible was built for a world where production was expensive. That world has changed.

What AI Changes

The capabilities that matter for instructional design practice are not the most dramatic capabilities AI has demonstrated. They are more mundane and, for that reason, more immediately consequential.

AI changes the economics of first-draft production. Scenario generation, objective drafting, assessment item creation, content outlining, feedback scripting: these are tasks that previously consumed significant designer time and can now be completed in a fraction of that time. This is a genuine change, not an incremental improvement. It shifts the designer's primary contribution from production to evaluation and judgment.

AI changes the cost of iteration. When a scenario can be regenerated in minutes rather than hours, the designer can afford to test more approaches, discard more early drafts, and refine more rigorously. The quality ceiling for a given time investment rises.

AI changes what synthesis work looks like. Stakeholder interviews, survey responses, performance data, facilitator debrief notes: these used to require manual reading and synthesis. AI can process and pattern-find across large datasets rapidly, which means the Enhancement phase described later in this book can be operated with a rigor that was previously impractical for most teams.

AI changes what a small team can produce. A solo practitioner with AI tools can now do analytical and production work that previously required a team. The capacity ceiling for L&D functions rises, regardless of size.

What AI Does Not Change

The judgment about whether a particular learning objective addresses the right performance gap is still a human judgment. AI can draft objectives fluently; it cannot determine whether the objective serves the learner's actual need.

The decision about whether training is the right solution to a performance problem is still a human decision. AI can generate training content for almost any topic; it cannot, on its own initiative, apply Gilbert's Behavior Engineering Model to determine that the problem is actually a feedback gap rather than a knowledge gap.

The assessment of whether a scenario will resonate as authentic for a specific audience in a specific organizational context is still a human assessment. AI can produce scenarios that are structurally sound; it cannot know whether the example reflects how the work actually happens in your organization.

The ethical responsibility for the content that learners encounter is still a human responsibility. Chapter 17 addresses this directly; the practitioner who uses AI tools and accepts their output without rigorous review has not accelerated quality. They have accelerated production of content of uncertain quality.

The relationship between a designer and the stakeholders who request their work, the facilitators who deliver it, the managers who champion it, and the learners who encounter it is still a human relationship. AI does not manage those relationships. It does not build the trust that makes L&D a strategic partner rather than a service function.

The core professional competency of instructional design has always been judgment: knowing what a learner needs, knowing when a design is working, knowing when to push back on a request that will not serve the people it is supposed to serve. That competency is not threatened by AI. It is made more necessary by it, because the volume of content that AI enables means the quality of the judgment applied to that content matters more, not less.

Introducing INDIE

INDIE is a framework for instructional design practice in an AI-augmented world. It stands for **IN**vestigation-Design, **D**evelopment, **I**mplementation, **E**nhancement. The hyphen in the first phase is intentional: analysis and design are concurrent streams that inform each other from the first client conversation, not sequential phases separated by a deliverable gate.

The framework is built on three premises. First, that the economic argument for sequential phases no longer holds in an AI-augmented workflow. Second, that the Enhancement phase is a continuous improvement cycle that begins at launch, not an evaluation activity that happens at the end. Third, that AI tools have specific strengths and specific limitations that must be understood clearly in order to use them well, and that using them well requires more professional judgment than using them poorly.

Chapter 5 presents the argument for INDIE in full, including an honest account of what ADDIE gets right and where it falls short. Chapters 6 through 9 address each INDIE phase in detail. Part Three addresses how the framework applies across the range of roles in the L&D ecosystem. The present chapter has established the conditions that make a new framework necessary. That context is the foundation on which everything that follows rests.

Maria, a Different Morning

Maria looked at the learning objective on her screen for another moment. Good objective. The question had not gone away: what exactly is her job now?

She had an answer, or the beginning of one. Her job was not to write objectives. Her job was to know which objectives were the right ones, at the right cognitive level, for these learners in this organizational context, in service of this performance outcome. The tool had produced the draft. She had evaluated it, and she had evaluated it correctly: it was good.

That evaluation was her job. That was what eleven years of practice had given her. The tool could produce the draft faster than she could. The tool could not do the evaluation, because the evaluation required knowing things the tool did not know: what the stakeholders had said that morning, what the performance data suggested about where the real gap was, what the learners in this organization needed that was different from what learners in a nominally similar organization needed.

She opened a new document and started working. Faster than before. With, she thought, more of her attention available for the things that actually required it.

Chapter Summary

- Instructional design's foundational theories remain valid. The process model built to deliver them, ADDIE, was designed for economic and technological conditions that no longer apply.
- ADDIE's sequential structure made sense when development was expensive and iteration was costly. AI changes both. The economic argument for sequential phases disappears when iteration becomes cheap.
- AI changes the economics of production, the cost of iteration, the scale of synthesis work, and the capacity ceiling for L&D teams of every size.
- AI does not change the professional judgments that define instructional design: which objectives address the right gap, whether training is the right solution, whether content is authentic for this audience, who bears ethical responsibility for what learners encounter.
- INDIE (INvestigation-Design, Development, Implementation, Enhancement) is the framework built for the conditions that now exist. The hyphen in the first phase reflects concurrent rather than sequential analysis and design. The Enhancement phase is continuous rather than terminal.

Reflection Questions

For individual journaling or team discussion.

1. Think about the last learning objective you wrote. How long did it take? What did that time contain? If you could produce the first draft in a fraction of that time, what would you do with the remainder? Would the objective be better or worse?
2. What is the most time-consuming production task in your current instructional design work? What would change about your role if that task took a tenth of the time it currently takes?
3. The chapter argues that AI makes the judgment component of instructional design more important, not less important. Do you agree? What is the professional judgment in your practice that AI cannot replicate, and how would you articulate it to a colleague who is anxious about the profession's future?
4. Look at your current process model. Where does it require sequential completion before the next phase can begin? What would it look like to make those phases concurrent rather than sequential?
5. Maria's question is "if I can do this in forty-five seconds, what exactly is my job now?" What is your answer to that question for your own practice?

Part One · Chapter 2

What AI Actually Is, and Isn't

James had been in learning and development for twelve years. He had a master's degree in instructional design, a shelf of well-annotated books on learning theory, and opinions about Bloom's Taxonomy that he was not shy to share. He also had a department head who had just returned from a leadership conference and kept using the phrase "AI-powered learning," four times in a single email. The email ended with a question: "Can we have a strategy by the end of the month?"

James stared at it. He had used AI tools, tentatively, curiously, sometimes skeptically. He had been impressed by some outputs and baffled by others. He did not feel qualified to build a strategy around something he didn't fully understand. But he also didn't know where to start understanding it. The explainers he found online seemed written for either software engineers or Machine Learning researchers. The books available were already out of date. The conference sessions were mostly vendors.

What he needed, he decided, was a plain account of what this technology actually was, what it could actually do, and where it actually fell short. Not marketing. Not alarm. Just an honest description. This chapter is that description.

The Honest Explanation

The AI tools that have entered instructional design practice over the past several years, the ones you use to draft content, analyze survey data, generate assessment items, produce images, and summarize research, are built on a class of technology called **L**arge **L**anguage **M**odels, or LLMs. Understanding what an LLM actually is, at a conceptual level, is the single most useful thing a practitioner can do to work with these tools intelligently. Here is the honest version, without the engineering detail you don't need and without the mystification that serves no one.

An LLM is a system trained on an enormous quantity of text, in the case of the most capable current models, something on the order of a significant fraction of the written internet, plus books, academic papers, code, and other structured text. At its core, the system is a machine learning model: a mathematical structure whose parameters are adjusted, through exposure to data, until the model becomes good at a specific task. For a language model, that task is prediction. Given any sequence of text, the system learns to predict what word, phrase, or sentence is likely to come next. It does this not by memorizing documents, but by finding the patterns and statistical relationships that run across hundreds of billions of examples, between words, concepts, sentences, and ideas. What makes this useful is that a model which has genuinely learned those patterns, rather than memorized specific instances, generalizes: it can produce coherent, contextually appropriate text for sequences it has never seen before. The result is a system that has internalized an extraordinarily rich representation of how human language works, what tends to follow what, and which ideas belong together, without having been explicitly told any of it. When you give an LLM a prompt, it generates a response by predicting, token by token, what text is most likely to follow your input, given everything it learned during training and everything you have told it in your conversation so far. It does this very, very quickly, and the results are often remarkably coherent, useful, and sometimes genuinely insightful. That is the mechanism. Now, here is what that mechanism means in practice, both the impressive parts and the important limitations.

What AI Is Genuinely Good At

The things AI tools do well are not random. They follow directly from the training process described above: an LLM is powerful precisely when the task involves recognizing patterns in language, generating structured text in response to a well-defined prompt, synthesizing information across a large body of text, or producing reasonable first drafts of things that have well-established forms.

For instructional designers, this means AI is a genuinely valuable partner in the following categories of work:

- **Drafting structured text.** Learning objectives written to Mager's format. Storyboard scripts following Gagné's events. Facilitation guides organized by session section. Assessment items in multiple formats. Job aid copy. Email communications to stakeholders. These are all tasks that have well-established structures, and LLMs have seen thousands of examples of each. The drafts they produce are often 70-80% of the way to usable, and getting them to 100% is significantly faster than starting from a blank page.

- **Pattern recognition in text data.** Open-ended survey responses. Interview transcripts. Performance review comments. Learner feedback collected over time. These are exactly the kinds of data that are valuable but expensive to analyze manually, and AI handles them well, surfacing themes, flagging outliers, and summarizing findings at a speed no human team can match. The analysis requires human verification, but the initial synthesis is genuine.

- **Generating variations and alternatives.** Need five different ways to open a module? Three versions of an assessment question at different cognitive levels? A scenario adapted for three different industries? AI produces variation quickly and without the fatigue that makes human writers default to their first instinct. This is particularly useful in assessment design, where cognitive diversity across items matters.

- **Explaining complex concepts accessibly.** Given a technical source document and a description of your audience, AI can produce a plain-language explanation of complex material with speed and reasonable accuracy. This is useful for content development when subject matter experts write at a level far above the learner audience, a common and persistent challenge.

- **Summarizing and synthesizing.** Long research reports, evaluation data sets, stakeholder interview collections; AI can condense these into usable summaries faster than any manual process. The summaries require expert review, but they provide a solid starting point that would otherwise take hours to produce.

- **Organizing and structuring information.** Curriculum maps, module outlines, competency frameworks, content taxonomies; AI can take a collection of ideas or content and suggest an organizational structure. It will not always suggest the right one, but it reliably suggests a reasonable one that can be evaluated and refined.

Where AI Reliably Fails

This is the part of the conversation that is least often heard clearly, because the people with the most to gain from your enthusiasm are not the people who will help you understand the limitations. So here they are, stated plainly, with specific relevance to instructional design practice.

It Confabulates

The technical term is "hallucination," but that word is too poetic. What actually happens is more specific and more important to understand: because an LLM generates text by predicting what should come next based on statistical patterns, it will sometimes generate text that is fluent, confident, and entirely fabricated. It doesn't know it's doing this. There is no internal alarm that fires when the model produces something false. The output simply looks like the output that would statistically follow from your prompt.

In instructional design, this shows up in specific ways. Ask AI to cite research supporting a learning theory claim, and it may produce citations to papers that do not exist, with correct-looking journal names, plausible author names, and realistic volume and issue numbers. Ask it about specific organizational compliance requirements, and it may produce something authoritative-sounding that is simply wrong. Ask it about the specific capabilities of a particular LMS or authoring tool, and it may describe features that the product doesn't have or has since changed.

The rule of thumb is simple and non-negotiable: *any AI output that makes a factual claim you will rely on for professional purposes must be independently verified.* This is not a limitation that will be fully resolved as models improve; it is structural to how current systems work. Build verification into your workflow, not as an afterthought but as a standard step.

It Has No Organizational Memory

An LLM knows nothing about your organization unless you tell it. It doesn't know your learners' actual experience levels, your company's culture, your stakeholders' personalities, the political history of the training program you're designing, or the specific language your organization uses for its competency frameworks. It will fill these gaps with plausible-sounding generic content, which is sometimes fine for a first draft, and sometimes quietly catastrophic when it gets into final materials.

The practical implication is that the quality of AI output in instructional design is directly proportional to the quality of the context you provide. A prompt that includes specific information about the learner audience, the organizational context, the content domain, and the design constraints will produce dramatically better output than a generic prompt. Building the habit of front-loading context into your prompts is one of the highest-return skills you can develop.

It Cannot Exercise Judgment

This is the most important limitation, and it is the one most obscured by the fluency of AI output. When an LLM produces a learning objective, it is not evaluating whether that objective addresses the right performance gap, whether it is appropriately ambitious for the learner audience, or whether it reflects a genuine organizational need. It is producing text that looks like a well-formed learning objective.

The difference matters enormously. An AI can draft five learning objectives in thirty seconds. Whether those objectives are the right five, whether they represent sound instructional design decisions grounded in a real understanding of the situation, is a question that requires human judgment. The same applies to needs analysis, curriculum architecture, assessment design, facilitation strategy, and every other dimension of instructional design work where expertise lives not in knowing the format but in knowing what the situation calls for.

It Has a Knowledge Cutoff

LLMs are trained on data collected up to a certain date, and they have no knowledge of events, research, or developments after that cutoff. For instructional designers, this matters in specific ways: AI may be unaware of updated compliance regulations, revised industry standards, recent research findings, new product versions, or organizational changes that have occurred since the model's training data was collected. In content areas that change rapidly, regulatory compliance, technology training, healthcare protocols, this is not a minor limitation. It requires systematic verification against current authoritative sources.

It Amplifies Bias

Because LLMs learn from human-generated text, and human-generated text reflects the assumptions, perspectives, and blind spots of the humans who wrote it, AI output can reproduce and sometimes amplify cultural, demographic, and linguistic biases in ways that are subtle and not immediately visible. Scenarios populated with default characters who reflect one demographic. Examples that assume a particular organizational culture. Language that is calibrated to one educational level or regional context without the model flagging that calibration.

Sidebar: The Six Most Common AI FAILURES in Instructional Design

Knowing the failure patterns in advance lets you design your workflow to catch them. Here are the six most common, with the specific risk each poses:

1. Fabricated citations and research claims
Risk: Content is published citing research that doesn't exist, or misattributing findings. Mitigation: Never use AI-generated citations without independently verifying each one in a database. If you cannot verify it, remove it.

2. Overconfident answers to domain-specific questions
Risk: AI produces specific, authoritative-sounding answers to questions about your industry, regulatory environment, or technical domain that are partially or entirely incorrect. Mitigation: For any domain-specific factual claim, route verification to a qualified SME or authoritative source.

3. Generic personas and audience profiles
Risk: AI-generated learner personas reflect a generic professional rather than your actual audience, leading to content misaligned with real learners' experience, language, and context. Mitigation: Always provide specific audience data in your prompt, and validate personas against actual stakeholder and learner input.

4. Homogeneous scenarios and examples
Risk: Default AI characters and scenarios cluster around a narrow demographic and cultural profile, producing materials that feel exclusionary or irrelevant to significant portions of the learner population. Mitigation: Ensure scenarios reflect your actual learner population and organizational context.

5. Objectives that look right but aren't
Risk: AI produces learning objectives in correct Mager format that nonetheless address the wrong performance level, miss the actual gap, or are unmeasurable in practice. The format is correct; the substance is not. Mitigation: Review every AI-generated objective against the actual needs analysis findings, not just against the format criteria.

6. Plausible-sounding but pedagogically weak assessments
Risk: AI assessment items often default to knowledge-recall questions even when higher cognitive levels are requested, and distractors in multiple-choice items are sometimes too obviously wrong. Mitigation: Review every assessment item against Bloom's level requirements and have a colleague evaluate distractor quality.

The Hype Problem, in Both Directions

There are currently two competing narratives about AI in learning and development, and both of them will lead you astray if you accept them uncritically.

The first narrative is triumphalist. AI will transform learning design. It will produce personalized experiences at scale, eliminate the need for most manual design work, and deliver measurable learning outcomes that traditional methods could never achieve. This narrative is popular at conferences, in vendor marketing, and in the emails of department heads who have just returned from conferences.

There is enough truth in this narrative to make it seductive. AI genuinely does transform certain aspects of design work. It genuinely does enable personalization that was previously impractical. It genuinely does accelerate processes that once took weeks. But the triumphalist version papers over the limitations described above, skips past the human expertise still required at every stage, and often conflates the capabilities of the most sophisticated research systems with the tools actually available to practitioners in real organizations with real budgets.

The second narrative is dismissive. AI often produces shallow, generic content. It cannot understand context, nuance, or organizational reality, for now anyway. It is a threat to professional quality and a shortcut that lazy designers will use to produce worse work faster. Experienced designers should resist it.

There is also enough truth in this narrative to make it seductive, particularly for practitioners who have seen genuinely poor AI-generated content, or who have watched colleagues use AI tools without the expertise to evaluate the output critically. But the dismissive version overstates the limitations, ignores the genuine capabilities described earlier in this chapter, and confuses the tool with how it's being used. A poor prompt that produces generic output is not evidence that AI cannot produce better; it is evidence that prompting is a skill.

The honest position is less exciting than either narrative. AI is a powerful set of tools that does certain things remarkably well, fails in specific and predictable ways, and requires skilled human judgment to use effectively. Its impact on instructional design practice is significant and real, but it does not eliminate the need for expertise; it changes what that expertise is applied to.

Two Roles AI Should Never Play

Given everything above, the most important practical distinction a designer can internalize is this: AI is a productive drafting and analysis partner, and it is a poor decision-maker.

As a drafting partner, AI accelerates the production of structured text, generates alternatives, synthesizes data, and handles the cognitive labor of getting from nothing to something. A designer who uses AI well in this role spends less time producing and more time evaluating, refining, and deciding. Their expertise goes further because the baseline work arrives faster.

As an analysis partner, AI can surface patterns in data that would take humans much longer to find, produce summaries that orient the designer to large bodies of information, and flag issues in content or design that benefit from a second look. Again, the human role is evaluation and judgment, the AI provides the raw material.

Where AI should not be positioned, and this bears stating firmly, is as a decision-maker. The question of whether training is the right intervention for a performance problem is a professional judgment call requiring knowledge of the organization, the people, and the systemic factors involved. The question of whether a particular instructional strategy is appropriate for a specific audience is a design decision that requires understanding of the learners and the context. The question of whether an assessment is measuring what it claims to measure is a validity question that requires psychometric expertise and situational awareness. None of these are questions AI can answer. All of them are questions that instructional designers exist to answer.

The risk of positioning AI as a decision-maker is not theoretical. It happens when a designer treats AI output as final without review. It happens when a manager instructs a team to 'use AI to do the needs analysis' without understanding what needs analysis actually involves.

It happens when an organization reduces its instructional design capacity because AI is handling the work, before anyone has checked what quality the AI is actually producing.

The most important thing you can do with the understanding in this chapter is to be clear, in your own practice and in your team's, about which role AI is playing at every stage of the work. Drafting partner. Analysis partner. Not decision-maker.

A Practical Mental Model

There is a mental model that has proven useful to many of the designers I have spoken with, and it is worth offering here. Think of an AI tool as a very well-read, very fast, very literal junior colleague who has never worked in your organization, has no personal judgment about what actually matters, and will confidently tell you something wrong if the pattern of your conversation points in that direction.

This colleague is genuinely useful. You would not want to be without them. They can draft things, research things, generate options, and handle the volume of structured work that would otherwise consume your whole week. But you would not send their output to a client without reviewing it. You would not trust their citation without checking it. You would not let them lead a stakeholder meeting. You would not ask them whether to redesign a program or cancel it.

You would work with them, deploy their strengths, compensate for their blind spots, and take responsibility for the decisions that are yours to make. That is the working relationship INDIE is built around. Not human versus machine. Not human replaced by machine. Human expertise, amplified and accelerated by machine capability, with clear professional judgment about what each brings and what each cannot.

What This Means for How INDIE Uses AI

Every AI application in the INDIE framework has been designed with the capabilities and limitations in this chapter in mind. You will notice that AI is consistently positioned as a generator of first drafts, a synthesizer of data, and a producer of options, never as the arbiter of design decisions. You will also notice that every phase includes explicit human review steps because the point at which the designer's judgment is applied to AI output is the point at which professional expertise is actually being exercised.

The prompts provided throughout the book are written to front-load context (addressing the organizational memory limitation), to request specific structures (reducing confabulation by anchoring the output in a defined form), and to explicitly invite the designer to evaluate and revise the output (reinforcing the drafting-partner relationship rather than the decision-maker one).

Understanding why these design choices were made will help you adapt the framework intelligently to situations the book didn't anticipate, which is, of course, most of your actual work.

AI Capability Reference for Instructional Designers

A practical guide to calibrating expectations across common ID tasks.

AI handles well ✓	Requires human judgment X
• Drafting learning objectives in Mager format	• Determining whether training is the right solution
• Generating scenario variations and case studies	• Interpreting politically sensitive stakeholder data
• Analyzing open-ended survey responses	• Making curriculum architecture decisions
• Writing facilitator guide sections	• Evaluating whether objectives address real gaps
• Producing assessment item drafts across formats	• Validating content accuracy in specialized domains
• Summarizing stakeholder interview themes	• Conducting root cause analysis of performance problems
• Generating job aid copy from structured content	• Assessing organizational readiness for a program
• Explaining technical content in plain language	• Navigating stakeholder conflict during design reviews
• Suggesting curriculum sequencing options	• Judging appropriate cognitive level for an audience
• Drafting stakeholder communications	• Making intervention selection decisions
• Producing Kirkpatrick Level 1 survey questions	• Evaluating transfer of learning to job performance
• Identifying structural gaps in draft content	• Determining ROI attribution for training programs

James, it turned out, did build the strategy. He sent it to his department head three weeks later, and when she asked how he had put it together so quickly, he gave her the honest answer: he had used AI tools for the drafting and the research synthesis, and he had applied his professional judgment to the decisions. He told her what the tools could do, and he told her what they couldn't. He told her which parts of the team's work were good candidates for AI augmentation, and which parts would always require human expertise, and why.

Chapter Summary

- Large language models generate text by predicting what should statistically follow a given input, a mechanism that explains both their capabilities and their failure modes.
- AI is genuinely strong at drafting structured text, recognizing patterns in qualitative data, generating variations, explaining complex material accessibly, and synthesizing large bodies of information.
- AI fails in specific, predictable ways: confabulation of facts and citations, absence of organizational context, inability to exercise professional judgment, knowledge cutoffs, and reproduction of bias from training data.
- The triumphalist narrative overstates AI's capabilities and understates the expertise required to use it well. The dismissive narrative overstates its limitations and confuses poor prompting with an inherent ceiling.
- The correct mental model: AI as a very fast, very well-read, very literal drafting and analysis partner who requires skilled human direction and critical review at every stage.
- AI should never be positioned as a decision-maker in instructional design work. The judgment calls that define our professional value, what to design, for whom, to what standard, measured how, remain irreducibly human.

Reflection Questions

For individual journaling or team discussion.

1. Think about the last time you used an AI tool at work. Which of the capability categories in this chapter does that task fall into? Did you treat the output as a draft to refine, or as a finished product? Would you make the same choice now?
2. Which of the six AI failure patterns in the sidebar do you consider the highest risk in your current work? What process change would reduce that risk most?
3. When you imagine explaining AI's capabilities and limitations to a skeptical colleague, what do you think their strongest objection would be? How would you address it honestly, without either overselling or dismissing?
4. Where in your current practice are you making decisions that you could describe as exercises of professional judgment, the kind AI cannot replicate? How clearly are those decisions visible to your stakeholders?
5. Where in your current practice are you making decisions that you could describe as exercises of professional judgment, the kind AI cannot replicate? How clearly are those decisions visible to your stakeholders?

Part One · Chapter 3

The Anxiety Is Legitimate, and Manageable

Diana had been an instructional designer for nineteen years. She had designed compliance programs, leadership development curricula, onboarding journeys, and technical training for industries ranging from healthcare to financial services. She had survived the eLearning transition, the mobile-first pivot, the LMS consolidation projects, and the microlearning mandate. She had earned a reputation as someone whose work was rigorous, thoughtful, and grounded, someone who pushed back on bad ideas and delivered good ones. She was also, in the spring of the year AI tools began arriving in her organization's workflow, quietly terrified.

Not loudly terrified. Quietly. She didn't say it in team meetings. She didn't say it to her manager. She maintained, in professional settings, the posture of someone who was thoughtfully evaluating new tools, assessing their fit, exercising due diligence. Inside, she was asking a question she didn't know how to answer: if a tool can do in twenty minutes what took me two days, then what, exactly, am I being paid for?

She didn't ask the question out loud because she was afraid of the answer. She was also afraid that asking it would make her look like someone who didn't know how to adapt, someone whose career had peaked, someone the organization would begin, quietly, to reconsider.

Diana is not a real person. But the experience she represents is real, and it is widespread, and it deserves more honest attention than it has received.

Let's Name What Is Actually Happening

The professional literature on AI in learning and development has produced a great deal of advice about opportunity, adaptation, and growth mindset. It has produced considerably less honest conversation about the human experience of watching a technology arrive that appears to threaten, or at least fundamentally complicate, the value of expertise that took years to build. So let's name what is actually happening for many experienced practitioners right now.

There is anxiety about obsolescence, the fear that AI will reduce demand for instructional designers in a way that previous technology shifts did not, because those shifts required designers while this one seems to partially replace them.

There is anxiety about competence, the discomfort of being a senior professional who does not fully understand a technology that junior colleagues seem to navigate more easily, and the attendant fear that this gap will become visible in ways that damage professional standing.

There is anxiety about quality, the genuine, professionally grounded concern that AI-augmented workflows will produce worse learning experiences, and that the pressure to adopt them will compromise the standards that define good instructional design.

There is anxiety about identity, the deeper, harder-to-articulate discomfort that comes when a professional's sense of self is built around the ability to do something well, and that something is suddenly being done by a machine.

Each of these is real. None of them should be dismissed with a confident reminder that humans will always be needed, or that this is just another technology shift, or that the key is to stay curious. Those things may be true, but they don't speak to the experience of sitting with the anxiety at nine o'clock on a Tuesday evening, wondering whether your career looks the way you thought it did.

This chapter will speak to those anxieties directly, not to dismiss them, but to examine them carefully enough to separate what is legitimate from what is distortion, and to find, in that distinction, something more useful than reassurance.

The question is not whether to feel the anxiety.
It is whether or not to let it make your decisions.

What History Actually Shows Us

Instructional design has not arrived at this moment unprepared. The profession has absorbed significant technological disruption before, and the pattern of how those disruptions unfolded is worth examining carefully, not because history always repeats, but because the pattern contains something important.

1990s	The eLearning Transition. The arrival of CD-ROM and web-based training created genuine anxiety among classroom facilitators and print-based designers. The concern: if content could be delivered by computer, what was the facilitator's role? Many predicted the end of instructor-led training. What actually happened: the volume of training required increased dramatically, the skill set expanded to include digital literacy and multimedia production, and the need for designers who understood both pedagogy and technology became more valuable, not less. The profession grew.

2000s	The LMS Revolution. Learning Management Systems moved the administrative and delivery infrastructure of training online. Designers who had managed programs through paper records, calendar systems, and direct coordinator relationships suddenly had to understand systems administration, data reporting, and the particular constraints of content developed for LMS delivery. The concern was about de-skilling, about training becoming more automated and less human. What actually happened: LMS implementation revealed how much design expertise was required to make these systems work for learners, and a new subspecialty of learning technology practice emerged.

2010s	Mobile, Microlearning, and the 'Moment of Need'. The shift to mobile-first design, five-minute modules, and performance support tools challenged designers whose work had been built around longer, classroom-based experiences. The concern: had learning been reduced to bullet points and push notifications? What actually happened: the field developed a more nuanced understanding of the different contexts in which learning occurs, formal and informal, and designers who understood how to match design decisions to context became more valuable than those who knew only one format.

Now	AI Augmentation. Tools that can draft, analyze, generate, and synthesize at machine speed have entered the design process itself. The concern: if AI can do what we do, faster and cheaper, what is the professional value of expertise? The question this chapter is designed to answer.

The pattern across all of these transitions is not that experienced practitioners were wrong to be concerned. In each case, the technology did change the work. Some roles were reduced or eliminated. Some organizations moved too fast and produced worse learning as a result. Real professional disruption occurred. The pattern is not that the anxiety was groundless.

The pattern is that the anxiety consistently overpredicted how much the technology would displace expertise, and underpredicted how much new expertise the technology would create demand for. It also consistently failed to anticipate the ways in which the technology, once absorbed into practice, actually elevated the importance of the judgment and relational skills that the technology could not provide. While that pattern certainly does not guarantee the same outcome this time around, it is evidence worth weighing.

Separating the Legitimate from the Distorted

Not all professional anxiety is equal. Some of it is a reasonable response to real risk, and it deserves to be taken seriously. Some of it is distortion, an amplification of risk driven by uncertainty, comparison, and the particular way that fear works on professional identity. Distinguishing between the two is not about minimizing the feeling. It is about responding to what is actually true rather than to what fear makes feel true.

The Legitimate Concerns

The concern about quality is legitimate. AI tools do produce shallow, generic, pedagogically weak output when used without expertise and critical review. Organizations that adopt AI-augmented workflows without investing in the design expertise required to use them well will produce worse learning experiences. This is already happening in some contexts, and experienced designers are right to name it and push back on it. The appropriate response is not to resist the tools; it is to establish and defend the quality standards that the tools must be held to.

The concern about organizational decision-making is legitimate. Some organizations will look at AI's drafting speed and conclude that they need fewer instructional designers. Some of those conclusions will be reached by people who do not understand what instructional designers actually do, and who are measuring the wrong things. This is a real risk, and it calls for a specific response: making the invisible work of design visible. The judgment calls, the stakeholder navigation, the pedagogical decision-making: these need to be articulated clearly to the people who make resourcing decisions, not assumed to be understood.

The concern about the pace of change is legitimate. The speed at which AI capabilities are developing is genuinely unprecedented in the history of tools available to our profession. What is true today about a tool's capabilities may not be true in six months. Staying current requires continuous learning investment that is real and takes real time. The appropriate response is to build that investment into your professional practice systematically, not to catch up in reactive bursts each time something changes.

The Distortions

The fear of obsolescence, in its most acute form, is largely a distortion, one rooted in a misunderstanding of what instructional design is. If the work of instructional design were primarily the production of structured text, writing objectives, drafting scripts, generating assessment items, then AI's speed would be an existential threat to the profession. But the production of structured text has never been what instructional design is actually for. It is an output of the work. The work itself is diagnosis, judgment, design, relationship, and advocacy. Those activities are not threatened by a tool that drafts text faster.

The anxiety about competence, specifically, the feeling that younger colleagues navigate AI tools more easily and that this represents a meaningful professional disadvantage, is worth examining carefully. It is true that people who grew up with digital interfaces have less friction with new digital tools. It is not true that ease of tool adoption is the same as quality of professional judgment. An early-career designer who adopts AI tools quickly but lacks the expertise to critically evaluate their output is not a stronger practitioner than a seasoned designer who adopts more slowly but evaluates rigorously. The direction of the advantage runs toward expertise, not toward tool fluency, especially as AI capabilities become table stakes rather than differentiators.

The anxiety about identity, the sense that expertise is being devalued, is perhaps the most understandable distortion, and also the most worth addressing directly. Professional identity built around the ability to produce things is vulnerable to tools that produce faster. Professional identity built around the ability to produce the right things, for the right reasons, with appropriate quality and professional judgment, that identity is not threatened by AI. It is clarified by it.

Sidebar: The Conversation You May Need to Have with Yourself

The following questions are not comfortable. They are offered in the spirit of the kind of honest self-examination that experienced professionals find more useful, ultimately, than reassurance. They are worth sitting with privately before answering if you answer them at all.

What specifically am I afraid will be taken from me?

Not in general terms, specifically. The answer is often more precise than the anxiety suggests. And the more precise it is, the more clearly you can assess whether the fear is proportionate to the actual risk.

Is my concern about AI, or is it about my organization?

Many practitioners whose anxiety is nominally about AI are actually anxious about how their organization handles change, whether leadership makes thoughtful decisions, whether expertise is valued, whether they personally have advocates. These are legitimate concerns, but they are organizational concerns, not technology concerns, and they call for different responses.

What would I need to believe about my own expertise to feel less threatened?

This question often reveals whether the anxiety is about the technology or about pre-existing uncertainty about professional identity. If the honest answer is "I would need to believe that what I do is actually valuable and not easily replaced," then the work is not primarily about AI.

What the Technology Cannot Do

The clearest antidote to the distorted version of professional anxiety is a precise account of what AI cannot do, not as reassurance, but as a factual description of where human expertise remains not merely preferable but irreplaceable. Chapter 2 began this accounting. This chapter completes it from the perspective of what the work of instructional design actually requires.

It Cannot Read the Room

Every experienced instructional designer knows that the work happens in rooms, physical and metaphorical, where what is said is only part of what matters. The stakeholder who agrees with the needs analysis but whose body language says otherwise. The subject matter expert who is technically cooperative but emotionally resistant to the training initiative. The department head who asked for compliance training but whose actual problem is a management issue that training cannot solve. The learner group who filled out the pre-assessment optimistically and whose actual skill level is significantly lower than the data suggested.

Reading these rooms, noticing what isn't being said, asking the question that unlocks the real problem, building the trust that allows a stakeholder to be honest about what they need, is the core of performance consulting and needs analysis. It is entirely human work. No current AI system can do it, and none on the near horizon will.

It Cannot Make Ethical Tradeoffs

Instructional design involves ethical decisions that are rarely named as such, but that nonetheless require moral reasoning. Whether to push back on a stakeholder who wants training used as a performance management tool rather than a genuine learning intervention. Whether to flag, at professional risk, that a mandated compliance program is pedagogically unsound. Whether to advocate for accessibility features that the budget doesn't include and the timeline doesn't accommodate. Whether to tell a client that their learning problem is actually a leadership problem.

These decisions require values, courage, and professional judgment. They are also the decisions that define the difference between an instructional designer and a content producer. AI can draft the materials either way. Only the designer can decide what the right thing is.

It Cannot Build Trust

The relationships that make instructional design work, with stakeholders who share real problems, with subject matter experts who give genuine access to their knowledge, with learners who engage honestly with assessments, with managers who implement transfer strategies because they believe in them, are all built on trust. Trust is built over time, through consistent behavior, demonstrated expertise, and human presence. It cannot be delegated to a tool. An experienced designer's most valuable professional asset is often not what they know or what they can produce. It is who trusts them and why. That asset is entirely immune to AI displacement.

It Cannot Hold the Standard

Someone has to decide what good learning design looks like. Someone has to push back when a course is pedagogically sound on paper and experientially deadening in practice. Someone has to ask whether the assessment is measuring what it claims to measure, whether the scenario reflects real learner experience, whether the feedback mechanism actually helps people improve. Someone has to care whether the learning works, not whether it was produced efficiently, but whether it actually produces competence in real people in real jobs.

That someone is the instructional designer. The presence of AI in the workflow does not change this. If anything, it makes it more important because AI produces output that can look like quality without being quality, and the gap between the two requires expert human discernment to find.

Three Practitioners, Three Different Paths Through

Abstract arguments about what AI can and cannot do are useful up to a point. What is more useful, for many people, is seeing how others have navigated the same transition. The following composites are drawn from conversations with practitioners at different career stages and organizational contexts. They are not success stories designed to inspire. They are honest accounts of how real professionals have found their way through a transition that is genuinely difficult.

The Veteran Who Reframed Her Value

Twenty-two years in the field. Deep expertise in healthcare compliance training. Skeptical of AI tools, not from ignorance but from professional standards. Her concern was specific: the content in her domain had to be precisely accurate, legally defensible, and clinically sound. She had seen what happened when those standards slipped. She was not interested in tools that produced plausible-sounding healthcare content at speed.

The shift for her came when she realized that her skepticism was itself a professional competency. She was the person in her organization who could evaluate AI output against clinical and regulatory standards, who could catch what the tool got wrong and articulate why it was wrong. She became the team's AI quality reviewer. The role required exactly the expertise she had spent two decades building. Her value increased.

She still thinks most AI-generated healthcare training content is inadequate. She is probably right about most of it. And she has found a way to make that judgment central to how she contributes.

The Mid-Career Designer Who Took the Risk

Twelve years in corporate learning, mostly financial services. Technically competent, professionally confident, but aware that her portfolio had become somewhat narrow. When AI tools arrived, she decided to lean in, not because she was told to, but because she recognized that getting ahead of the transition was better than being caught by it.

She spent three months experimenting systematically: using AI tools across every phase of projects, documenting what worked and what didn't, building a personal library of prompts that she refined through iteration. She made mistakes. She produced some things that went to a colleague for review and came back with significant corrections. She got faster, and then she got better. Eighteen months later, she describes the transition as the most significant professional investment she has made. Not because AI made her work easier, it changed the nature of the work in ways that required learning, not just adjustment, but because leaning into the discomfort early meant she was competent before it became expected. The window for when you can get credit for figuring it out is shorter than people think.

The New Designer Who Found Her Footing

Three years in the profession. Entered instructional design just as AI tools were becoming widely available. Her anxiety was different from her more experienced colleagues': not that AI would replace her, but that she would never develop the expertise that made experienced designers valuable, because AI would short-circuit the slow, difficult, formative work of learning to design.

She had a point. There is something in the early years of instructional design practice that happens through struggle, through writing a bad objective and having a mentor explain why it fails, through designing an assessment that doesn't measure what you thought it measured and having to figure out why. The danger of AI for early-career designers is not replacement but shortcut: producing adequate-looking work without developing the judgment to know what makes it adequate.

Her response was deliberate: she used AI tools, but she required herself to evaluate every output against the theoretical frameworks she was learning. Not "does this look right?" but "does this meet Mager's criteria, and if not, why not, and what would it take to fix it?" The AI became a sparring partner rather than a producer. She got more reps in a shorter time, which, used this way, may actually accelerate expertise development rather than impede it.

Moving Through It

The three practitioners above found different paths because they needed different things. There is no single right way to navigate this transition. But across many conversations with designers at every career stage, a few consistent principles have emerged, not as a recipe, but as an orientation.

- **Name what you're actually afraid of.** Generalized anxiety is hard to address because it has no specific target. The more precisely you can name what you fear, job loss, quality decline, competence gap, identity disruption, the more clearly you can assess what it would actually take to address it, and whether the fear is proportionate to the risk.

- **Start with something specific, not something comprehensive.** The practitioners who adapted most effectively did not set out to transform their entire workflow. They chose one task, writing learning objectives, analyzing survey data, drafting assessment items, and learned to do it with AI tools before expanding. Comprehensive change is paralyzing. Specific experiments are manageable.

- **Build evaluation into the habit from the beginning.** The difference between a practitioner who uses AI well and one who uses it poorly is not tool fluency. It is the habit of critical review. Every AI output should be evaluated against a professional standard, not accepted as finished. This habit is easier to build at the beginning of a new practice than to retrofit later.

- **Make your judgment visible.** If the work AI cannot do is the exercise of professional judgment, then that judgment needs to be visible to stakeholders, to managers, to the organization. Document the decisions you make and why. Articulate the tradeoffs in design reviews. Name the ethical considerations in stakeholder conversations. Don't assume that because you know the value of what you're contributing, others can see it.

- **Find the people who are doing this seriously.** Every professional community contains people who are engaging with this transition thoughtfully, not marketing it, not fearing it, but working through it with rigor and honesty. Finding those people, inside or outside your organization, is one of the highest-return things you can do. This is precisely where Vygotsky's insight about social learning is most applicable: we think better, and learn faster, in genuine community with others who are working on the same problems.

- **Allow yourself to not have it figured out.** The practitioners who appeared most competent about AI in the early days of this transition were often the least trustworthy guides, because they had resolved their uncertainty through confidence rather than through understanding. The honest answer to 'how do you think about AI in your practice?' is not a polished framework. It is a set of principles you are still developing, experiments you are still running, and questions you are still sitting with. That is not a weakness. It is what learning looks like.

A Word About Diana

Diana, the designer who opened this chapter, quietly terrified, is still at her organization. She still hasn't said the word "terrified" out loud in a professional setting. But she has said something closer to honest: that she finds the pace of change uncomfortable, that she has concerns about quality, and that she is working through how her expertise applies in a workflow that looks different from the one she built it in.

Her manager, who is better than most, heard this and gave her something useful: a specific project, a specific AI tool, and permission to do it slowly. Diana spent six weeks on what might have been a two-week project, evaluating every output carefully and documenting what the tool did well and what it missed. She came out of it with something she hadn't expected: not confidence that AI was good, but confidence that she could tell when it was not.

That, it turned out, was what she had been afraid of losing, her ability to know. And she still had it. She just had to find it in a new context.

The anxiety didn't go away. But it became manageable, which is what the title of this chapter promises and all that honesty can deliver. The goal is not to feel certain in an uncertain moment. It is to move through the uncertainty with your professional judgment intact, which is the same goal that good instructional design has always set for its learners.

Chapter Summary

- The professional anxiety that AI generates among experienced instructional designers is real, widespread, and deserves honest attention rather than dismissal.
- The anxiety takes four main forms: concern about obsolescence, about competence, about quality, and about professional identity. Each deserves a specific, honest response.
- History shows that previous technology shifts, eLearning, LMS, mobile, consistently overpredicted displacement and underpredicted the new demand for expertise. The same pattern is likely, though not guaranteed, with AI.
- Some professional concerns are legitimate: quality risk when AI tools are used without expertise, organizational decisions based on misunderstanding of design work, and the real pace of required learning. These call for specific, active responses.
- Some concerns are distortions: the fear of obsolescence assumes that text production is what instructional design is for. The fear of competence disadvantage confuses tool fluency with professional judgment. The fear of identity loss is real but not proportionate.
- What AI genuinely cannot do, read rooms, make ethical tradeoffs, build trust, hold the standard of quality, is precisely the work that defines instructional design's professional value.
- Moving through the anxiety productively means naming it specifically, starting with small experiments, building evaluation habits from the start, making professional judgment visible, and finding genuine community with others navigating the same transition.

Reflection Questions

For individual journaling or team discussion.

1. Which of the four anxiety types, obsolescence, competence, quality, identity, resonates most with your own experience right now? What would an honest, specific response to that particular anxiety look like?
2. Think about the work you have done in the past month that required professional judgment that AI could not have provided. How visible was that judgment to the people who observed or evaluated your work?
3. Of the three practitioner composites in this chapter, the veteran, the mid-career designer, the new practitioner, whose situation is closest to yours? What did their path through suggest that might be useful in your own?
4. If a junior colleague came to you today and said they were anxious about AI, what would you tell them? Does the advice you would give them apply to you as well?
5. What is the one specific experiment, one task, one tool, one project, that would move you from theoretical understanding of AI-augmented design to genuine practical experience? What is preventing you from running that experiment?

Part One · Chapter 4
Working with AI: Prompt Engineering and Context Architecture

Sofia had been using AI tools for eight months and the results were driving her quietly mad. Not because the tools were bad. She had seen them do things that genuinely impressed her: draft a scenario in sixty seconds that would have taken her an hour, synthesize three stakeholder interviews into a coherent theme map, produce a set of learning objectives that were structurally sound and needed only modest revision. She had shown these outputs to colleagues, and the reaction was uniformly the same: enthusiasm that stopped just short of awe.

But the results were inconsistent in ways she could not explain. The same prompt that produced a sharp, usable needs analysis on Tuesday produced a generic, vague one on Thursday. A scenario generation prompt that worked beautifully for one program produced scenarios for another that felt disconnected from the actual work her learners did: technically correct, professionally hollow. She had tried refining the prompt language. She had tried being more specific. She had tried different phrasing. The results were still inconsistent, and she could not locate the variable.

The conversation that finally clarified it happened at a peer learning session, when a colleague described how she prepared materials before starting any AI-assisted analysis. She described building what she called a "project brief," a structured document that assembled everything the AI would need to reason intelligently about this specific project, this specific audience, this specific organizational context. She provided this document at the start of every session.

Sofia listened and felt the particular sensation of understanding something she should have understood much earlier. The prompt was not the variable. The context was.

The Fundamental Misunderstanding

Most practitioners who begin using AI tools invest their improvement effort in the prompt: the exact wording, the specific instructions, the format of the request. This is not wrong, prompt quality matters, but it addresses the less important of the two variables that determine AI output quality. The more important variable is the context: the information the AI has access to when it generates a response.

This distinction is consequential enough to state precisely. A language model does not have opinions. It does not have knowledge of your organization, your learners, your subject matter domain, or your project history. What it has is a very large amount of training data and a sophisticated ability to generate text that is statistically consistent with that data. When you give it a prompt without substantive context, it generates a response that is statistically consistent with how that prompt is typically answered across all the contexts it has seen, which is to say, it generates a generic response. Not a bad response, necessarily. A generic one.

The generic response is the source of every inconsistency Sofia was experiencing. When her Tuesday prompt worked and her Thursday prompt did not, the difference was not in the prompt. It was in what she had given the model to work with. On Tuesday she had included, perhaps without noticing, a specific detail about the audience or the organizational context that anchored the model's response to something real. On Thursday she had not.

The practitioner who understands this shifts their investment accordingly: less time refining prompt language, more time building the context that makes the prompt's output genuinely useful. The prompt tells the AI what to do. The context tells it what to do it with. The context is almost always the binding constraint.

> *The prompt tells the AI what to do. The context tells it what to do it with. The context is almost always the binding constraint.*

What RAG Is: The Practitioner's Version

Retrieval-Augmented Generation (RAG) is the technical name for a straightforward idea: before generating a response, the AI retrieves relevant information from a curated set of your documents and then generates its response with that material in hand.

Here is the plain-language version. Imagine you are a brand-new instructional designer on your first day. Someone asks you to write a knowledge check about your company's lockout/tagout procedures. You have two options. Option one: answer from whatever you remember from your college courses and general industry knowledge. Option two: walk to the file cabinet, pull out your company's actual lockout/tagout procedure document, read the relevant sections, and then write the knowledge check based on what you just read.

RAG is option two, automated for an AI model. The "retrieval" happens in seconds, and the "augmented generation" means the response is anchored to your source material rather than floating on general knowledge. In a regulated industry like the energy sector, this grounding is not a luxury. It is a requirement.

Why General AI Falls Short for L&D Work

Large language models like Claude, Llama, and the model behind Copilot (the major platforms available to practitioners at the time of writing, though the specific tools will continue to evolve faster than the principles for using them) are trained on enormous amounts of publicly available text. They know a great deal about the world in general. But they do not know your company's specific procedures, your facility's equipment configurations, your regulatory commitments, or the particular way your organization structures its training programs. When you ask a general AI model about these topics, it will often produce something that sounds right but is actually wrong in ways that matter, wrong procedure step numbers, outdated regulatory references, or generic safety language that does not match your approved protocols. In the energy sector, this gap is not merely inconvenient. Training content that misrepresents a procedure, omits a safety step, or cites a superseded regulation can lead to real operational consequences. RAG closes this gap by ensuring the AI's output is rooted in your approved, current source material.

For instructional designers using AI tools in a standard chat interface, RAG in its practical form is this: before submitting any analytically demanding prompt, assemble the relevant project documentation into the conversation context. This is not a technical operation. It is a professional one. It requires knowing what material is relevant to this query, how to present it clearly, and how to structure it so the model can reason against it effectively. These are design skills, not engineering skills.

Key Terms

The following terms will appear throughout this chapter and throughout the book.

Term	What It Means for Your Work
LLM (Large Language Model)	The AI engine you interact with when you use Claude, Llama-based tools, or Copilot. It generates text based on patterns it learned during training, plus whatever context you provide in your prompt.
RAG	Retrieval-Augmented Generation. A method of giving the AI access to specific documents before it generates a response, so the output is grounded in your actual source material rather than the model's general training.
Knowledge Base	The curated collection of documents you want the AI to draw from. In an L&D context, this could include SOPs, regulatory guidance, technical manuals, lesson plans, approved training content, and stakeholder interview notes.
Embedding	A behind-the-scenes process that converts text into numerical codes representing its meaning. This is how the system identifies which documents are relevant to your question. You do not need to create embeddings manually — the tools handle this.
Vector Database	A specialized storage system that holds embeddings and can quickly find the most relevant documents for a given query. Think of it as a very fast, meaning-aware filing system.
Chunk	A segment of a larger document. Instead of feeding an entire 200-page manual to the AI, the system breaks it into smaller pieces so it can retrieve just the parts relevant to your question.
Context Window	The maximum amount of text the AI can "see" and work with at one time. Even large context windows have limits, which is

	why RAG retrieves only the most relevant content rather than entire document libraries.
Hallucination/Confabulation	When an AI generates information that is plausible-sounding but factually wrong. RAG dramatically reduces hallucination by giving the AI real source material to reference. It does not eliminate it.
Grounding	The practice of tying AI-generated content back to verified source material so you can confirm its accuracy. Every AI output used in learning content should be grounded before publication.
Project Knowledge Base	A living document or folder that assembles everything the AI needs to reason intelligently about a specific project: stakeholder notes, performance data, source documents, design decisions, and Enhancement data.

How RAG Works: A Practitioner's Walkthrough

You do not need to understand the engineering behind RAG in order to use it effectively, but having a working knowledge of the process will help you troubleshoot problems, communicate with technical colleagues, and make better decisions about how to structure your knowledge base and your prompts.

Part 1: Preparing Your Documents

Before a RAG system can retrieve your documents, they need to be processed and organized. In dedicated enterprise RAG systems, this is handled automatically. In the chat-interface tools most practitioners use, it is handled by you when you prepare and upload your source material.

1. **Collect.** Gather the source documents you want the AI to reference. For L&D work, this means SOPs, work instructions, regulatory guidance, technical manuals, previous course content, safety bulletins, stakeholder interview notes, performance data, and any other material relevant to the project at hand.
2. **Clean.** Ensure documents are in a machine-readable format. Scanned PDFs may need optical character recognition (OCR) before they are useful. Remove watermarks, headers, footers, and other artifacts that add noise without adding content.
3. **Organize.** Use descriptive filenames and consistent folder structure. "SOP-4.7.2_Confined_Space_Entry_Rev3_2025.pdf" is more useful than "Document1.pdf." The AI uses filenames as

context clues. Consistent labeling makes your knowledge base navigable for both you and the model.

4. **Prioritize quality over quantity.** A focused set of 20 high-quality, well-organized documents will produce better results than 200 randomly assembled ones. Every document in your knowledge base should earn its place.

Part 2: Asking a Question

When you submit a query through a RAG-enabled tool, the following happens in seconds:

1. **Your question is embedded.** The system converts your question into the same type of numerical representation used for the document chunks.
2. **Relevant content is retrieved.** The system compares your question against all stored document content and finds the closest matches, the sections whose meaning is most similar to your question.
3. **A prompt is assembled.** The retrieved content is combined with your original question and any instructions you have provided into a single prompt sent to the LLM.
4. **The AI generates a response.** The LLM reads the retrieved content and your question, then generates a response that draws on the specific source material rather than general training patterns.
5. **You receive a grounded answer.** The response is based on your actual documents. Your job is then to evaluate it against the source and apply the expert judgment the AI cannot provide.

The Critical Insight

The quality of a RAG workflow depends on two things you directly control: the quality of the documents in the knowledge base and the quality of the prompts you write. The AI platform handles the technical plumbing, but you determine what goes into it and how it is used. The AI handles the retrieval and text generation, but you provide the three things it cannot: judgment about which source material is authoritative and current, expertise in how to structure effective learning experiences, and the final quality check that ensures every piece of training content is accurate, complete, and safe.

How This Works Across the Tools You Use

Claude, Microsoft Copilot, and Llama-based tools each implement RAG-like functionality in different ways. Understanding these differences will help you choose the right tool for each task and configure it effectively.

Platform	How It Handles RAG	Best For	Limitations
Claude (Anthropic)	Upload documents directly into conversations or Projects. Claude reads and references them in its responses. Projects allow persistent document collections across multiple conversations.	Deep analysis, long documents, nuanced writing tasks, complex procedure-based content development.	Documents must be uploaded per conversation or organized in Projects. No permanent enterprise-wide knowledge base without API integration.
Copilot (Microsoft)	Integrates with Microsoft 365. Can reference SharePoint files, emails, Teams chats, and uploaded documents. Searches across organizational content automatically.	Summarizing internal documents already in SharePoint or OneDrive, drafting in Word or PowerPoint with reference to existing organizational files.	Retrieval quality depends on Microsoft Graph indexing. Less control over which documents are retrieved. May surface outdated or irrelevant content from SharePoint.
Llama (Meta)	Open-source models deployable within custom RAG pipelines. Typically accessed through tools your IT team provisions or through third-party interfaces supporting document upload.	Organizations needing on-premises AI for security or compliance reasons. Customizable retrieval pipelines.	Requires more technical setup than Claude or Copilot. You will likely depend on your IT team or a developer to configure the RAG pipeline.

When Context Matters Most

Not all AI tasks in instructional design have the same context dependency. Understanding which tasks require rich context and which can function adequately without it is the first practical skill of AI-augmented practice.

Thin Context: Generic Output	Rich Context: Specific Output
Generating a generic scenario about a customer service interaction	Generating a scenario that reflects how your specific organization handles account escalation and what has gone wrong in the past
Drafting a learning objective in Mager format for a topic	Drafting a learning objective that addresses the specific performance gap your needs analysis identified in this population
Producing a Kirkpatrick four-level analysis framework	Analyzing actual Level 1 through Level 4 data from the last three cohorts of this specific program and identifying patterns
Writing a competency description for a role	Auditing a competency description against the job task analysis you conducted and the SME knowledge you elicited last month
Generating stakeholder interview questions for a training project	Generating targeted hypothesis-testing questions for your specific root cause hypothesis about why error rates rose in the northeast region
Producing a curriculum architecture for a safety training program	Designing a sequenced curriculum that builds on the existing onboarding program, reflects the time constraints established in the L&D manager conversation, and addresses the gaps the persona research identified

The left column describes tasks where generic output is often sufficient because the task itself is generic: drafting a format, producing a template, illustrating a principle. The right column describes tasks where generic output fails because the task is inherently specific: analyzing this data, diagnosing this gap, designing for this audience. Almost all of the analytically consequential tasks in instructional design are in the right column.

The prompts in this book that address right-column tasks contain placeholder fields in brackets: [PASTE STAKEHOLDER INTERVIEW NOTES], [DESCRIBE YOUR HYPOTHESIS], [PASTE LEVEL 2 DATA BY OBJECTIVE]. These placeholders are not formatting conventions. They are the context requirement made explicit. Leaving them unfilled and submitting the prompt anyway is the single most common error in AI-augmented instructional design practice. It produces a generic response that looks like analysis and is not.

Building Your Project Knowledge Base

The most productive structural change a practitioner can make in their AI-augmented workflow is building and maintaining a project knowledge base: a living document or folder that assembles the information the AI needs to reason intelligently about this specific project, updated as the project progresses and referenced at the start of every analytically demanding session.

The project knowledge base is not a summary of the project. It is a structured collection of the source material that the AI can reason against. The distinction matters: a summary tells the AI what you concluded. The source material lets the AI reach its own conclusions: and, more usefully, lets it challenge yours. A model that has access to the raw stakeholder interview transcript can notice a dissonance you edited out of your summary. A model working from your summary cannot.

What to Include

Document Type	What to Include · How to Structure · When to Update
Stakeholder Notes or Transcripts	**Include:** Full notes or verbatim transcripts of all stakeholder conversations, including informal conversations that contained relevant information. Do not summarize before including, include the source. **Structure:** Label each document with the stakeholder name, role, and date. If multiple conversations occurred, preserve each separately rather than merging them. The AI will synthesize; your job is to provide the raw material. **Update when:** Each time a new stakeholder conversation occurs. Older conversations remain in the base; they provide the longitudinal record the AI needs to notice when positions have shifted.

Performance and Organizational Data	**Include:** Error rate data, assessment scores, production metrics, completion rates, engagement survey results, and any other quantitative data that describes the performance gap or organizational context. **Structure:** Include the data with enough contextual labeling that the AI understands what it represents: "Q3 error rate by team, baseline period Jan–Mar, measurement period Apr-Jun." Numbers without context produce analysis without meaning. **Update when:** When new data cycles become available. Include historical data even if it predates the current project, patterns across time are often more informative than point-in-time snapshots.

Source Procedures and Regulatory Documents	**Include:** Current, approved versions of SOPs, work instructions, regulatory guidance, equipment manuals, and technical standards relevant to the performance domain. Include full documents rather than excerpts where possible. **Structure:** Pull documents from your organization's document management system (Documentum, SharePoint, or similar) to ensure you have the current approved version. Use descriptive filenames including document type, subject, version, and date. **Update when:** When documents are updated. Flag superseded versions clearly rather than deleting them, knowing what used to be true and when it changed is often analytically important.

Learner Analysis Materials	**Include:** Survey results, focus group notes, observation records, and any other primary research on the learner population. Include both quantitative results and qualitative verbatim responses. **Structure:** Keep open-ended responses verbatim. The AI cannot analyze sentiment and subtext in summarized language; it can in original language. The verbatim record is the analytical asset. **Update when:** After each learner research activity. The knowledge base is most valuable when it accumulates across the IN phase, what you learned in week one shapes how you interpret what you learn in week three.

Design Decisions and Rationale	**Include:** A running record of significant design decisions: what was decided, why, what alternatives were considered, and what information the decision was based on. **Structure:** Informal, dated log entries. The AI needs to know what has already been decided and why in order to produce recommendations consistent with the project's direction rather than starting from zero each session. **Update when:** After every significant design conversation or decision point. The log is most valuable when it is current enough to represent the project's actual state.

Enhancement Data	**Include:** Level 1 through Level 4 data as it accumulates across delivery cycles: survey results, assessment performance by objective, facilitator debrief notes, transfer observation data, and business metric records.

	Structure: Organized by delivery cycle and Kirkpatrick level. Each cycle should be a labeled section so the AI can track patterns across cycles rather than analyzing each in isolation. **Update when:** After each delivery cycle. The power of AI-assisted Enhancement analysis is pattern recognition across multiple cycles; that pattern recognition is only possible when all cycle data is present.

What to Exclude

- Superseded or draft versions of documents (unless clearly marked and needed for comparison)
- Documents outside your area of responsibility or expertise
- Content you do not have authorization to use or redistribute
- Low-quality scans or documents with significant OCR errors that the AI cannot read reliably
- Redundant copies of the same document
- Sensitive information that should not be processed by external AI tools, review your organization's AI use policy before including any personal data, security-classified details, or information governed by specific regulatory constraints

How to Organize for AI Use

- **Label everything with sufficient context to stand alone.** Every document, data table, and note in the knowledge base should be readable by someone who has no other context about the project. This means including organization name (or a consistent anonymized substitute), dates, roles, and the relationship of the document to the project. 'Survey results' is less useful than "Level 1 survey results, Module 2 pilot cohort, March delivery, 18 participants, financial services division."
- **Separate source material from conclusions.** The knowledge base should contain the raw material from which conclusions are drawn, not only the conclusions themselves. If it contains only your synthesis of stakeholder interviews rather than the interviews themselves, the AI can help you develop the synthesis but cannot challenge it. If it contains the interviews, the AI can both develop and challenge the synthesis.

- **Maintain a current-state summary at the top.** At the start of the knowledge base, maintain a brief (one page or less) current state summary: what phase the project is in, what has been decided, what questions are currently open, and what the next decision point is. This summary is what the AI reads first; it orients all subsequent reasoning. An outdated current state summary is worse than no summary.
- **One version, one truth.** Never have multiple versions of the same document in your knowledge base simultaneously. When a procedure is revised, replace the old version entirely. Simultaneous versions confuse the model and produce contradictory outputs.

Document Preparation Checklist

Before adding any document to your project knowledge base:

□ Verify the document is the current approved version

□ Confirm text is selectable and readable (not a scanned image without OCR)

□ Apply a descriptive filename including document type, subject, version, and date

□ Remove or redact sensitive information not needed for training content development

□ For very long documents (100+ pages), consider splitting into logical, self-contained sections

□ Convert spreadsheets or tables with complex formatting to plain text or CSV before uploading

□ Note the document's title, version, date, and source in your tracking log

Platform-Specific Implementation

The specific approach to building and using your knowledge base varies by platform. The following workflows address the three tools most commonly used by L&D practitioners in the energy sector.

Using Claude (Anthropic)

Claude offers two primary approaches for working with your documents, and understanding which to use when is the first practical skill of working with Claude effectively.

- **Conversation uploads.** Attach documents directly to a conversation using the file attachment feature. Claude will read and reference these documents within that conversation. This is ideal for one-time tasks or quick questions. Upload the documents at the start of the conversation before asking your questions.
- **Projects.** Create a Project in Claude and upload your document set to it. Every conversation within that Project will have access to these documents. This is the better approach for ongoing work – create a Project for each course you are developing, each regulatory area, or each job role's training program.

> ▶ **Claude Project Setup for Course Development**
>
> Create a Project titled with the course name or program area. Upload all relevant source documents. Then add a custom instruction in the Project's system prompt area such as: "You are an assistant for instructional designers developing energy sector training. Base all responses on the uploaded documents. When referencing procedures, cite the specific document name, section number, and step. If the documents do not contain information to answer a question, say so. Never fabricate procedure steps or regulatory requirements." This instruction anchors every conversation in the Project to your specific source material.

Using Microsoft Copilot

Microsoft 365 Copilot automatically has access to content across your organization's Microsoft 365 environment, SharePoint, OneDrive, Teams, and Outlook. This means your RAG knowledge base may already partially exist in your organization's document libraries.

- **Leverage existing SharePoint structure.** If your team's procedure documents, training materials, and regulatory references are already stored in SharePoint, Copilot can search across them. Ensure the documents are in libraries you have access to and that they are properly tagged with metadata.
- **Reference specific files.** In Copilot chat, reference specific files by name or use the "/" command to point Copilot at particular documents or SharePoint sites. This narrows the retrieval to the

most relevant sources rather than allowing Copilot to pull from the entire organizational content corpus.

- **Use Copilot in Word.** When drafting training documents, use Copilot within Word and reference your source files. This keeps your writing grounded in the source material while you work in a familiar environment. Verify all content against the original documents before finalizing.

Using Llama-Based Tools

If your organization has deployed Llama models through an internal platform or a third-party tool, the document upload process will depend on how that platform is configured. Common approaches include web-based chat interfaces that allow document uploads similar to Claude, API-connected tools where your IT team has built a custom interface with document ingestion capabilities, and integration platforms like LangChain or LlamaIndex.

If you are unsure how to upload documents to your Llama-based tool, coordinate with your IT team or the platform administrator. Provide them with this chapter's section on knowledge base preparation so they understand how you need documents organized and what you are trying to accomplish. You are the expert on what L&D needs from these tools; they are the experts on how to configure them. Both perspectives are required.

Prompt Engineering: Seven Principles

With a well-constructed project knowledge base in place, the prompt itself becomes the refinement layer rather than the primary determinant of output quality. Good prompt engineering is not about finding magic words. It is about structuring requests so the AI can reason precisely about the specific task at hand.

Principle 1: Provide context before instruction
The AI should understand the situation before it receives the task. Begin any analytically demanding prompt by establishing who the learner is, what the organizational context is, what has already been established, and what question is currently open. The instruction, what you want the AI to do, follows this orientation.

Example: Instead of "Write three learning objectives for a compliance training program," begin with "The following program addresses a specific gap in our vendor management process where audits have identified inconsistent documentation practices. Objectives must address the judgment calls, not just the procedures. Now write three learning objectives..."

Principle 2: Assign a specific perspective

Language models are more useful when given an explicit perspective to occupy than when asked to respond from a neutral position. The perspective does not have to be a role title; it can be a specific orientation, skeptical reviewer, experienced practitioner in this domain, stakeholder whose concerns have not been addressed.

Example: Instead of "Review this learning objective," try "Read this learning objective as a skeptical performance consultant who has seen many compliance programs fail to change behavior. What questions would you raise?" The assigned perspective changes both what the AI attends to and how it frames its response.

Principle 3: Specify the output format before the output

Tell the AI what structure the response should take before asking for the content. This forces the AI to organize its reasoning in ways that serve your specific use of the output. An unformatted response requires you to extract the structure afterward. A pre-specified format requires the AI to apply that structure during generation, which produces more disciplined analysis.

Example: Instead of "Analyze this feedback data," specify: "Organize your analysis into four sections: (1) patterns across all respondents, (2) outliers that contradict the pattern, (3) questions the data does not answer, (4) the single most important finding. Write in prose, not bullet points." The format specification produces a response you can use directly.

Principle 4: Make the constraint explicit

The most useful AI outputs are often produced under explicit constraints that mirror the real conditions of the work: this program has six hours of learning time, this audience has no prior experience with the regulatory framework, this objective must be assessable in a 15-minute eLearning module. Constraints that exist in your professional context but are left unstated produce outputs that are theoretically sound and practically unusable.

Example: Compare "Design a curriculum for safety training" with "Design a curriculum for safety training given: 4-hour maximum total learning time, blended delivery with a 2-hour ILT and 2 hours of eLearning, audience of experienced technicians who resist content-heavy instruction, LMS that supports xAPI but not branching scenarios above 3 levels." The second version produces a design you can actually build.

Principle 5: Ask for the reasoning, not just the conclusion

For analytical tasks, the reasoning process is often more valuable than the conclusion it produces. A model that explains why it classified a root cause as primarily motivational rather than knowledge-based is a model whose reasoning you can evaluate and correct. Ask for reasoning explicitly, particularly for any output that will inform a significant design or management decision.

Example: Add to any analytical prompt: "For each conclusion you reach, explain the specific evidence from the provided materials that supports it, and identify the strongest evidence that could challenge it." This converts an output that feels authoritative into one you can reason with.

Principle 6: Iterate rather than escalate

The instinct when AI output is inadequate is to write a longer, more complex prompt. This is usually the wrong response. A better response is to identify the specific element that failed and address that element directly in a follow-up prompt. The context established in earlier turns is an asset; abandon it only when the direction of the session is fundamentally wrong.

Example: If a scenario output lacks organizational authenticity, do not rewrite the entire prompt. Instead: "The scenario is technically accurate but does not reflect how this work actually happens in an energy sector context. Here are three specific ways the scenario fails that test: [LIST]. Revise the scenario to address these three failures specifically." The targeted correction produces better results than a wholesale restart.

Principle 7: Name what you are uncertain about

Explicitly identifying the element of your own thinking that you are least certain about produces outputs that address that uncertainty directly, rather than reinforcing the parts of your thinking you are already confident in. The practitioner who presents their work as settled receives confirmation. The practitioner who names their doubt receives challenge.

Example: Add to any design review prompt: "The element of this design I am least confident about is [NAME IT SPECIFICALLY]. Please read the design specifically for weaknesses in that area, and tell me whether my concern is justified and what I should do about it."

The Anatomy of a Good RAG Prompt

The seven principles above address the decisions that matter most in prompt design. For analytical tasks against specific documents, an effective prompt has four required components. Any prompt missing one of these four will produce a less useful output than one that includes all of them.

- **Role and context.** Tell the AI who it is and what it is helping with. Example: "You are an instructional design assistant helping me develop a training module on electrical safety for field technicians at an energy sector organization. Base all responses exclusively on the uploaded documents."
- **Task.** Clearly state what you want. Be specific. "Generate 10 multiple-choice questions" is better than "make some quiz questions." "Identify procedural steps in the uploaded SOP that are not covered in the current training module" is better than "compare these documents."
- **Source instructions.** Tell the AI how to use the documents. 'Base each question on specific content from the uploaded procedure documents. Include the source document name and section number after each question. If a question cannot be grounded in the source documents, do not include it.' This last sentence is critical; it is your guardrail against hallucination.
- **Output format.** Specify the format you need. "Each question should include four answer options (A–D), the correct answer, a brief rationale explaining why it is correct, and a reference to the source document." Format specification prevents the AI from making its own layout decisions that may not serve your use case.

Prompt Examples for Common L&D Tasks

The following examples illustrate the four-component structure applied to the tasks instructional designers most frequently need AI assistance with. These are starting point structures; adapt them to your specific project context and uploaded documents.

Task	Example Prompt Structure
Draft Learning Objectives	Review the uploaded procedure document for Section 4.3 (Turbine Shutdown). Write 4–6 terminal learning objectives for a training module covering this procedure. Each objective should use a measurable action verb at the Application or Analysis level of Bloom's Taxonomy and reference the specific procedural steps. Format: Given [condition], the learner will [action verb] [criterion].
Generate Assessment Items	Based on the uploaded SOP for confined space entry, create 8 multiple-choice questions suitable for an annual requalification exam. Each question must test application or analysis-level understanding (not just recall). Include 4 options with one correct answer. Cite the specific SOP section each question addresses. Do not include information not found in the uploaded document.
Gap Analysis	Compare the content in the uploaded training module against the uploaded revision of SOP-2.4.1. Identify any procedural steps, warnings, or requirements in the current SOP that are not covered or are inaccurately represented in the training module. Present findings in a table with columns: SOP Section, SOP Requirement, Training Coverage (covered/missing/inaccurate), Recommended Action.
Regulatory Summary	Summarize the key compliance requirements from the uploaded OSHA 1910.146 guidance document as they apply to our confined space entry training program. Organize by topic area. For each requirement, note whether our current training (in the uploaded lesson plan) addresses it. If the documents do not contain enough information to answer a question, say so rather than inferring.
Job Aid Creation	Using the uploaded equipment maintenance procedure, create a one-page quick-reference job aid for field technicians performing quarterly inspections. Include only the critical steps, safety precautions, and required PPE. Use simple, direct language appropriate for a field reference document. Cite the source procedure section for each item.
Content Review	Review the uploaded draft training module for technical accuracy against the uploaded source procedures. Flag any statements not supported by the source documents, any safety-critical information that appears to be missing, and any areas where the training language could be misinterpreted. Provide specific recommendations for each finding with the source document and section that should govern the correction.

Prompting Mistakes to Avoid

- **Being vague.** "Make some training content" gives the AI no useful direction. Specify the topic, audience, format, length, and source requirements.
- **Forgetting to require citations.** If you do not explicitly ask the AI to cite its sources, it will often blend document content with general knowledge, making it impossible to verify accuracy. Always require citations in a consistent format: "After each factual

claim, note the source in brackets: [Document Name, Section X.X]."

- **Uploading too many unrelated documents.** A focused set of relevant documents produces better results than a large, unfocused collection. The AI may retrieve irrelevant chunks from unrelated documents and blend them into your response.
- **Skipping the uncertainty instruction.** Always include a line like "If the documents do not contain enough information to answer this, say so rather than guessing." This is your guardrail against hallucination on the specific content that matters most.
- **Not specifying the audience.** Content written for a new hire is very different from content written for an experienced operator. Always tell the AI who the learners are, what they already know, and what experience they bring.

Evaluating RAG Output: The FACTS Framework

Every response from a RAG-enabled AI should be treated as a draft that requires human review. Your expertise as an L&D professional is the quality gate. The FACTS framework provides a structured approach to evaluating each AI-generated output before using it in any training material

Criterion	Question to Ask	What to Look For
F: Faithful	Is every factual claim traceable to a specific source document?	Check cited sections. Flag any claims that do not appear in the source material.
A: Accurate	Are the facts, figures, procedure steps, and regulatory references correct?	Cross-reference critical details against the original documents, especially step sequences and safety warnings.
C: Complete	Does the output cover everything it needs to, without critical gaps?	Compare against the source to identify missing steps, omitted precautions, or incomplete explanations.
T: Targeted	Is the content appropriate for the intended audience and learning level?	Check vocabulary, complexity, assumed prior knowledge, and alignment with stated learning objectives.

S: Safe	Could any part of this content, if followed, create a safety risk?	Pay special attention to procedure sequences, PPE requirements, warnings, cautions, and hazard descriptions.

The Non-Negotiable Rule for Safety-Critical Content

In the energy sector, AI-generated content involving safety procedures, regulatory requirements, or operational sequences must always be reviewed by a qualified subject matter expert before it is used in any training deliverable. RAG improves accuracy significantly, but it does not eliminate the need for human verification on safety-critical content. A confident-sounding AI response about a lockout/tagout procedure that misrepresents one step is more dangerous than an obviously incomplete response, because it may pass a surface review without triggering SME scrutiny. The SME review is non-negotiable. Build it into the project timeline.

Verify Regulatory References Independently

When the AI cites a specific regulation (e.g., "OSHA 1910.147(c)(4)(i)"), verify the citation against the actual regulation before including it in any training deliverable. AI models can generate plausible but incorrect regulatory citation formats. The section number, subsection letter, and specific requirement language all require independent verification. A wrong citation that looks right is worse than no citation, because it may be relied upon without checking.

Common Problems and Solutions

The following table addresses the issues practitioners most frequently encounter when working with RAG-enabled AI tools, with practical solutions you can implement immediately.

Problem	Likely Cause	Solution
The AI ignores my documents and answers from general knowledge	Your prompt does not clearly instruct the AI to prioritize uploaded documents, or the documents were not properly uploaded or recognized.	Add explicit instructions: "Answer based only on the uploaded documents." Verify documents uploaded successfully. In Claude, check that files appear in the conversation or Project before submitting your prompt.

The AI cites the wrong document or wrong section	The knowledge base contains multiple documents with similar content, or document chunks are too large for precise retrieval.	Use more descriptive file names. In your prompt, specify which document(s) to prioritize. Consider splitting large documents into focused, self-contained sections before uploading.
Responses are generic and not specific enough	The prompt lacks specificity about the audience, context, or desired level of detail.	Add more context: specify the learner audience, the training context, the level of detail needed, and provide an example of the desired output format. Review Principles 1 and 4.
The AI produces plausible but incorrect procedure steps	Hallucination blending retrieved content with general knowledge. This is the most dangerous failure mode in energy sector training.	Strengthen your grounding instruction: "Do not infer or reconstruct procedure steps. Only state steps that are explicitly written in the source document." Always verify procedure content with a qualified SME.
Tables, lists, or structured content from documents is garbled	Complex formatting (tables, multi-column layouts, nested lists) can be lost during document processing.	Pre-process complex tables into plain-text descriptions before uploading. For critical structured content, copy and paste the table content directly into your prompt rather than relying on document upload.
The AI seems to forget uploaded content mid-conversation	Very long conversations or extremely large document sets can approach the AI's context window limit.	Start new conversations for new tasks. Keep document sets focused on the immediate task. In Claude, use Projects to maintain document access across shorter, focused conversations.
Different team members get inconsistent results	Variation in prompting style, document organization, or platform settings across users.	Standardize with shared prompt templates, a common document preparation checklist, and consistent Project configurations. Hold a team calibration session where everyone runs the same prompt against the same document and compares outputs.

I cannot tell what came from my documents versus general AI knowledge	The AI blends sources without clear attribution.	Require explicit citations in every prompt: "For each statement, indicate in brackets whether it comes from [document name] or from your general knowledge." If a response cannot be attributed to a document, treat it as unverified.

Quick-Start: Your First RAG Workflow

Use the following checklist to set up your first RAG-enabled workflow. The entire process can be completed in a single working session of approximately one hour. Choose Claude for your first attempt, its document upload and Projects features are the most straightforward for practitioners new to RAG workflows.

Before You Start (15 minutes)

□ Choose one specific, repeatable task from your current workload (e.g., drafting quiz questions from a procedure)

□ Identify 3–5 source documents that are essential for that task

□ Verify all documents are current, approved versions from your document control system

□ Apply descriptive file names using the convention: DocumentType_Subject_RevVersion_Date

Setup (20 minutes)

□ Create a Claude Project (or equivalent workspace) with a clear, descriptive name

□ Upload your selected documents to the Project

□ Write a custom instruction that defines the AI's role, grounds it in your documents, and instructs it to acknowledge uncertainty rather than fabricate

□ Verify the AI can see the documents by asking a simple factual question about their content

First Test (15 minutes)

□ Write a detailed prompt using the four-component structure: Role and context, Task, Source instructions, Output format

□ Submit the prompt and review the output using the FACTS framework

□ Note what worked and what needs improvement

□ For any safety or regulatory content, flag for SME review before use

Refinement (10 minutes)

- □ Adjust your prompt based on the first test results
- □ Resubmit and compare the improved output
- □ Save your refined prompt as a template for future use
- □ Share results with a colleague to validate quality

Scale (Ongoing)

- □ Expand to additional use cases as you build confidence
- □ Add documents to your knowledge base as new material becomes available
- □ Build your team's shared prompt library with tested, effective prompts
- □ Schedule quarterly knowledge base reviews to remove outdated content
- □ Document lessons learned and share across your work unit

The Prompts in This Book as Learning Exercises

Every prompt in this book was designed with the preceding principles in mind. Each prompt is a worked example of a specific prompt engineering approach applied to a specific instructional design task. But using the prompts as templates, filling in the brackets and submitting, produces only a fraction of their value. The full value comes from understanding why each prompt is structured the way it is, and from developing the ability to construct prompts of equivalent quality for the tasks that the book's prompts do not cover.

With that goal in mind, the following approach to the prompts in subsequent chapters is recommended:

- **Before using a prompt, analyze its structure.** Identify: what context does the prompt establish before issuing the instruction? What perspective does it assign? What constraints does it specify? What format does it request? What reasoning does it ask for? Understanding the structure is more valuable than memorizing the language.
- **Use the prompt with the richest context you can assemble.** Every bracket is a context invitation. The difference between a prompt submitted with a one-sentence stakeholder description

and the same prompt submitted with the full interview transcript is the difference between a generic response and a useful one. Invest in the context, not in refining the prompt language.

- **Apply the FACTS framework to every output before using it.** This is non-negotiable for any content that will be delivered to learners. For safety-critical or regulatory content, add SME review as a fifth step beyond the FACTS evaluation you can perform yourself.
- **Develop your own variants for your specific context.** No prompt in this book was written for your specific organization, your specific learner population, or your specific domain. The prompt that is most valuable to your practice is one you have adapted from a book prompt or constructed from scratch, tested against real project materials, refined based on the results, and documented in your team's shared library. That prompt represents accumulated professional knowledge. A copied template represents a starting point.

Sofia's Workflow, Revised

The change Sofia made after the peer learning session was not dramatic. She did not adopt a new tool or rebuild her entire workflow. She started maintaining a project brief document for each active project, a structured file she updated at the end of each work session, organized into sections: what she knew about the learners, what the stakeholder conversations had established, what the current design decisions were and why, what questions remained open. She pasted the relevant sections into the conversation context before any analytically demanding prompt.

The inconsistency disappeared. Not because she had found better prompt language, she had not significantly changed her prompts, but because she had given the model something consistent to reason against. The Tuesday-Thursday variation had never been about the model. It had been about the context she was providing, which varied based on what she happened to remember to include on any given day. The project brief made the context consistent, session to session, regardless of what she happened to remember.

She also started noticing what the AI was doing with the context she provided, which was more interesting than she had expected. It was not simply retrieving and reporting her materials back to her. It was finding the dissonances between them: the stakeholder who said the problem was knowledge when the performance data suggested it was incentives, the objective she had written at the analysis level when the job required the evaluation level, the scenario she had approved that reflected how the procedure was documented rather than how it was actually performed.

That, the AI as a second reader of her own project materials, noticing what she was too close to see, turned out to be the most valuable thing it did. Not because it was smarter than she was. Because it was not her, and it was reading what was there rather than what she meant to put there.

The difference, she understood now, had never been in the tool.

Chapter Summary

- The most consequential variable in AI output quality is not the prompt language but the context: the information the model has access to when generating its response. Improving prompts without improving context produces marginally better generic outputs, not genuinely useful specific ones.
- RAG (Retrieval-Augmented Generation) is the method of giving the AI access to specific documents before it generates a response. In practical terms for L&D professionals, it means assembling relevant project documentation into the conversation context before submitting analytically demanding prompts.
- In the energy sector, RAG is not a convenience, it is a quality requirement. Training content built from general AI knowledge may misrepresent specific procedures, cite superseded regulations, or produce plausible-sounding safety guidance that does not match your approved protocols.
- The project knowledge base, a structured, maintained collection of stakeholder notes, performance data, source procedures, learner research, design decision logs, and Enhancement data, is the foundation that makes AI-assisted work genuinely project-specific rather than generically plausible.
- The seven principles of prompt engineering address the decisions that matter most: context before instruction, assigned perspective, pre-specified

output format, explicit constraints, reasoning alongside conclusions, iteration over escalation, and naming what you are uncertain about.

- The FACTS framework, Faithful, Accurate, Complete, Targeted, Safe, provides a structured evaluation approach for every AI-generated output before it enters any training deliverable. Safety-critical content requires SME review in addition to FACTS evaluation. This is non-negotiable.
- The prompts in this book are worked examples of prompt engineering applied to instructional design tasks. Analyze their structure, use them with rich context, apply FACTS to their outputs, and develop your own variants suited to your specific practice. Build the team's shared prompt library from what you learn.

Reflection Questions

For individual journaling or team discussion.

1. Think about the last time you used an AI tool and were disappointed by the output. Looking back through the lens of this chapter: was the problem the prompt language, or the context you provided? What would you include in the context now that you did not include then?
2. Apply the specificity test to the last three AI outputs you accepted and used in a project: could each have been generated for a different project, a different audience, a different organization? If yes, what context would have made them specific rather than generic?
3. What does your current project workflow capture that could form the basis of a project knowledge base? What is not currently captured in any document that you carry in your head and would need to externalize for the knowledge base to function?
4. Apply the FACTS framework to a piece of AI-generated content you have used in a training deliverable in the past three months. Is it Faithful, Accurate, Complete, Targeted, and Safe? If you cannot verify all five criteria, what would verification require?

Part Two · Chapter 5

Why ADDIE Isn't Enough, and What INDIE Offers

The whiteboard in the conference room had the familiar shape on it. Five boxes in a row, connected by arrows: Analysis, Design, Development, Implementation, Evaluation. Someone had written the acronym in large letters above it: ADDIE. Below it, in smaller letters: Our Process.

Marcus, who had joined the L&D team eight months ago after a decade in instructional design consulting, looked at it and said nothing for a moment. The team was meeting to discuss a new program: a leadership development curriculum for newly promoted managers, to be launched in ten weeks. Ten weeks. Five phases.

He did the arithmetic quietly. Four weeks of analysis and design before anything got built. Two weeks of development, if the content review went smoothly, which it never did. Implementation in week seven, which meant piloting in week eight at the earliest. Evaluation after that, which meant the first meaningful data on whether the program worked would arrive after the launch. He had been in this room before, in other organizations, with other whiteboards, doing the same arithmetic and arriving at the same conclusion: the model doesn't fit the timeline, but we'll use it anyway, and the timeline will be significantly overshot. "We should talk about the process," he said. His manager looked up. "We use ADDIE." "I know," Marcus said. "We should talk about the process."

The Honest Case Against ADDIE

Critiquing ADDIE feels, in some quarters of the instructional design profession, like speaking ill of a grandparent. The model has been the profession's foundational framework for more than fifty years. Many of the designers reading this book were trained in it, have used it throughout their careers, and have defended it in conversations with stakeholders who wanted to skip straight to development. There is genuine affection for it, and that affection is not irrational.

But affection is not the same as fitness for purpose, and honesty about a model's limitations is not the same as disrespect for what it accomplished. ADDIE gave the instructional design profession a shared language, a defensible process, and a systematic approach that elevated the quality of training work in contexts where it was applied well. That is a real legacy. It is also a legacy that belongs to a different era of design practice. Here is the honest case for why ADDIE is no longer sufficient as the primary framework for contemporary instructional design work, and why the limitations are structural, not incidental.

The Linear Assumption

ADDIE was built on an assumption that made sense in the 1970s: that the phases of instructional design are genuinely sequential, and that completing each phase thoroughly before moving to the next produces better outcomes than working on them simultaneously. Analysis yields stable requirements. Stable requirements enable clean design. Clean design drives predictable development. Predictable development supports confident implementation.

The problem is that this is not how instructional design actually works in practice, and it has not been for some time. Analysis surfaces requirements that the design process reveals are incomplete. Design decisions surface questions that require additional analysis. Development exposes content gaps that require design revision. Implementation reveals learner needs that were not visible in the analysis. Every experienced designer knows this. ADDIE as practiced has always been more iterative than ADDIE as described, which means the model describes a process that designers have already had to depart from in order to do good work.

A framework that requires practitioners to work around it in order to succeed is not serving the profession. It is serving the appearance of the profession.

The Analysis-Design Fiction

The separation of Analysis from Design in ADDIE's model is its most consequential structural decision, and the one most directly challenged by contemporary practice. The assumption is that you can fully understand a learning problem before you begin to design solutions, that analysis and design are genuinely separable activities with a clean handoff between them.

They are not. They never were, if we are being precise. Any designer who has conducted a thorough needs assessment knows that the process of asking good analytical questions requires you to be thinking about design possibilities, because the questions worth asking depend on what kinds of solutions are available. And any designer who has begun designing a course knows that design decisions immediately generate analytical questions about learners, contexts, and constraints that the front-end analysis left unanswered.

AI has made this fiction impossible to maintain with a straight face. When an AI-powered analytics platform can simultaneously identify a learning gap in performance data and generate a draft curriculum structure to address it, the phase boundary between Analysis and Design is not just blurred. It is obsolete. The tools have caught up to what experienced designers have always known: these activities inform each other constantly, and separating them creates waste without improving quality.

Evaluation at the End

Placing Evaluation as the final phase in ADDIE's sequence was a reasonable design choice given the technology of its era. If data on learning effectiveness can only be collected after a program has been delivered, then evaluation is necessarily a post-implementation activity. The model was built around the constraints that existed when it was created.

Those constraints no longer exist. Modern learning management systems generate continuous data on learner engagement, progress, and assessment performance. Survey tools can collect and analyze reaction data in real time. Performance analytics can identify behavior change indicators within weeks of program completion. The data infrastructure for evaluation now runs alongside implementation rather than after it, but ADDIE's structure still treats evaluation as something that happens when everything else is done.

The practical consequence is that organizations using ADDIE's evaluation model in a data-rich environment are systematically underusing the information available to them. They collect data they could act on and treat it as material for a report rather than a signal for improvement. This is not a quality problem. It is a structural problem, and it requires a structural solution.

The Speed-to-Value Gap

The most immediately felt limitation of ADDIE in modern organizations is simpler than any of the above: it is too slow. Not because its practitioners move slowly, but because its architecture assumes development timelines measured in months and delivers results on timelines that organizations increasingly cannot wait for.

The half-life of workplace skills is compressing. Competitive pressures demand faster responses to learning needs. AI tools have already reduced the production time for many development tasks significantly. But if the design architecture requires four weeks of analysis and formal signoff before a single piece of content is built, the speed gains from AI-augmented development are partially consumed by a front-end process that hasn't changed.

INDIE doesn't solve this by cutting corners. It solves it by running activities that ADDIE sequences in parallel, and by building the feedback loops that allow early, smaller launches to surface information that would otherwise require a full development cycle to discover.

> *ADDIE as practiced has always been more iterative than ADDIE as described. A framework that requires practitioners to work around it is not serving the profession.*

What INDIE Is, and What it Isn't

Before presenting the framework, it is worth being precise about what INDIE does and does not claim to be.

INDIE is not a wholesale rejection of systematic instructional design. The principles that gave ADDIE its staying power, understand learner needs rigorously, write clear and measurable objectives, design assessments that reflect real performance, measure effectiveness at multiple levels, are fully present in INDIE. They are not optional additions; they are the foundation on which INDIE is built. Practitioners who have internalized ADDIE's discipline will find that INDIE gives structure to practices they have already been developing informally, not a new set of principles to learn from scratch.

INDIE is not an agile framework for software development adapted for learning design, though it draws on agile methodology's emphasis on iteration and rapid feedback. The differences between software development and learning design are real and consequential. Learning outcomes are not user stories. Learner experience is not a product interface. The frameworks inform each other without being interchangeable.

INDIE is not the only legitimate approach to instructional design in an AI-augmented environment. Organizations with specific compliance requirements, highly regulated content, or institutional needs for formal phase completion may have good reasons to maintain a more structured process. The comparison table at the end of this chapter maps the situations in which each model is the better fit.

What INDIE is: a contemporary framework for instructional design practice that takes the capabilities and constraints of modern AI tools seriously, structures the work to take advantage of continuous data rather than treating data as a post-implementation resource, and describes how learning design actually happens in competent, well-resourced teams rather than how it is supposed to happen according to a model developed before the current technological landscape existed.

The Four Phases: A First Look

Each phase of INDIE is addressed in its own chapter in the pages that follow. What this section offers is the logic of the framework, why the phases are structured as they are, what each is designed to accomplish, and how they relate to each other. Think of it as the map before the territory.

IN	**INvestigation-Design** *The fusion that changes everything* Analysis and design are not sequential activities; they are parallel, mutually informing streams that belong in the same phase. Investigation encompasses needs assessment, learner analysis, performance gap identification, and stakeholder engagement. Design encompasses objective development, instructional strategy selection, curriculum architecture, and assessment planning. In INDIE, these streams run concurrently, with AI tools enabling real-time synthesis of analytical data into design recommendations, and design decisions surfacing new analytical questions as naturally as they always have.
D	**Development** *Building with intelligence* Content creation, system configuration, and quality assurance, accelerated by AI and structured by Gagné's Nine Events of Instruction. Development in INDIE is not just faster than traditional development; it is more systematically grounded, because the AI prompts that drive content creation are built around the psychological conditions for learning rather than around content outlines. Human expertise remains essential throughout: to evaluate AI output, to make pedagogical decisions that require contextual judgment, and to hold the quality standard that AI cannot hold for itself.
I	**Implementation** *Agile deployment for real organizations* INDIE replaces the single, high-stakes launch with iterative deployment: pilot with a representative group, collect and analyze feedback with AI tools, adjust in near-real-time, then roll out more broadly. This approach reduces launch risk, accelerates the discovery of what actually works with real learners in real contexts, and creates a culture of improvement from the first day of delivery rather than after a formal evaluation cycle. Stakeholder management, facilitation support, and project coordination are all addressed in this phase, with AI handling the drafting and synthesis work that frees designers for the relational and judgment-intensive activities that define implementation quality.
E	**Enhancement** *From evaluation to continuous improvement* The most significant philosophical shift in INDIE: replacing the end-phase evaluation checkpoint with a continuous enhancement process that runs in parallel with implementation and extends indefinitely beyond it. Enhancement encompasses everything Kirkpatrick's Four Levels capture: reaction, learning, behavior change, and business results, but it treats them as an ongoing information stream rather than a summative report. AI provides the analytical capacity to process that stream continuously. Human designers provide the judgment to act on what the data reveals. Learning solutions become adaptive ecosystems rather than static artifacts.

The Relationship Between the Phases

One of the most important things to understand about INDIE is what happens between the phases, not just within them. The arrows in ADDIE's diagram point in one direction. The relationships in INDIE point in multiple directions simultaneously, because the framework is designed around information flow rather than phase completion.

Investigation-Design feeds Development with objectives, strategies, and content architecture. Development feeds Implementation with materials, facilitation guides, and assessment instruments. Implementation feeds Enhancement with delivery data, learner performance, and facilitator observations. And Enhancement feeds back into all three preceding phases: informing the next iteration of Investigation-Design with data on what worked and what didn't, triggering Development revisions based on learner performance patterns, and refining Implementation strategies based on delivery experience.

This is not just a philosophical position about iteration. It is a practical description of how well-functioning learning programs actually evolve over time in organizations that treat learning as an ongoing investment rather than a series of discrete production projects. INDIE gives that evolution a structure.

It also means that INDIE is not a process you complete and then start over from scratch. It is a cycle that a program enters and moves through continuously, with each pass through the cycle producing a more refined and more effective learning experience. The first pass looks most like a traditional design project. The second and subsequent passes look more like product management: data-informed, iterative, and continuously improving.

ADDIE vs. INDIE: A Structured Comparison

The table below maps both frameworks across eight key dimensions. It is not intended as a verdict; the final row identifies contexts in which ADDIE remains the appropriate choice. It is intended as a clear account of the differences, so practitioners can make informed decisions about which framework fits which situation.

Dimension	ADDIE	INDIE
Phase structure	Five discrete, sequential phases with defined handoffs between each.	Four integrated phases, with Investigation and Design operating as concurrent streams rather than sequential steps.
Analysis / Design relationship	Analysis must be complete before Design begins. The two phases are separated by a formal deliverable boundary.	Investigation and Design are fused into a single phase, acknowledging that understanding needs and generating solutions are simultaneous, mutually informing activities.
Iteration	Iteration is possible in principle but structurally discouraged, returning to an earlier phase requires formal scope change.	Iteration is the default. Each phase is designed to feed information back into the others, and pilot data shapes development decisions.
AI integration	Not addressed. The model predates AI augmentation of design work and provides no guidance on how to incorporate it.	AI is integrated at every phase, with specific applications, prompt frameworks, and human review requirements designed in from the start.
Evaluation philosophy	Evaluation is the final phase, a retrospective check on whether the program met its objectives. Improvement follows in a subsequent project cycle.	Enhancement replaces Evaluation, reframing improvement as a continuous, data-driven process that runs alongside and beyond implementation rather than after it.
Deployment approach	A single, comprehensive launch following the completion of development. Pilot testing occurs within the Implementation phase but typically precedes full rollout.	Iterative deployment with rapid feedback cycles. Pilot programs launch quickly, data is collected and acted on in near-real-time, and full rollout follows demonstrated effectiveness.
Theoretical grounding	Implicitly grounded in systematic design principles; does not explicitly integrate contemporary learning theory frameworks into its process guidance.	Explicitly grounded in Knowles, Vygotsky, Gagné, Piaget, Mager, and Kirkpatrick at each phase, with prompts and checkpoints that operationalize these frameworks.
Appropriate context	Stable content domains, long development cycles, well-defined requirements, lower tolerance for iteration, and contexts where formal phase completion is organizationally important.	Dynamic content domains, compressed timelines, data-rich environments, AI-augmented teams, and contexts where iterative improvement and responsiveness are organizational priorities.

When ADDIE Is Still the Right Choice

A framework that claims to be appropriate for every context is probably appropriate for none of them. INDIE is not universally the better choice, and practitioners who adopt it should be clear about the contexts in which ADDIE's structure still serves a legitimate purpose.

Highly regulated content domains, nuclear safety, clinical procedure, aviation, may require the formal phase gates and documentation checkpoints that ADDIE's structure provides. In these contexts, the ability to demonstrate that a rigorous analysis phase was completed and formally signed off before development began is not a bureaucratic inconvenience. It is a professional and sometimes legal requirement.

Organizations with institutional requirements for formal deliverable review at each phase, certain government agencies, large public sector organizations, and some academic institutions, may have governance structures that ADDIE was designed to accommodate and that INDIE's iterative approach is harder to fit within. In these contexts, the practical path forward is often to use INDIE's practices within ADDIE's documentation structure: running Investigation and Design concurrently while still producing the formal Analysis deliverable that the governance process requires.

Programs with extremely stable, long-lived content in domains where learner needs and organizational context are well understood and unlikely to change significantly, certain foundational technical certifications, for example, may genuinely benefit from ADDIE's thoroughness more than from INDIE's emphasis on adaptability. If the learning solution will not need to change, the infrastructure for continuous improvement is overhead rather than value.

The honest answer is that most contemporary instructional design work falls outside these exceptions; it involves content that changes, timelines that compress, data that is available for analysis, and learner needs that require responsive design. For that work, INDIE is the more appropriate framework. Knowing when the exception applies is itself an exercise of the professional judgment that this book is trying to develop.

Back to Marcus

The conversation in the conference room lasted forty minutes. Marcus did not dismantle the whiteboard or redraw the diagram. He did something simpler: he described what the ten-week timeline actually required, and he proposed a process that fit it.

The needs assessment and the first draft of the learning objectives would happen in the same week, not sequentially. Stakeholder interviews and content design sessions would be combined, not separated. A pilot with twelve participants would launch in week six, before development was technically complete, with the final two modules revised based on what the pilot revealed. Evaluation data would be collected from day one of the pilot, not after the full program launched.

His manager listened, asked two clarifying questions, and said: "This isn't how we normally do it." "No," Marcus said. "But it's how we'll actually meet the deadline. And it's how we'll know whether what we built worked before we've fully committed to it." There was a pause. His manager looked at the whiteboard. Then she picked up the marker and drew a curved arrow from Implementation back to Design. "Show me how this works," she said. The next five chapters are the answer to that question.

Chapter Summary

- ADDIE's limitations are structural, not incidental: its linear architecture, the artificial separation of Analysis from Design, the placement of Evaluation at the end, and its architecture's incompatibility with compressed timelines are all baked into the model's design.
- ADDIE as practiced has always required iteration that ADDIE as described does not accommodate, meaning experienced designers have been working around the model in order to do good work. INDIE formalizes what they have already been doing.
- INDIE is built on ADDIE's systematic foundation, not against it. The principles that gave ADDIE its value are fully present; what changes is the architecture that organizes them.
- The four phases, Investigation-Design, Development, Implementation, and Enhancement, are designed around information flow and iteration rather than phase completion and sequential handoffs.

- INDIE's phases are not just structurally different from ADDIE's. They relate to each other differently: Enhancement feeds back into all three preceding phases continuously, making INDIE a cycle rather than a sequence.
- ADDIE remains the appropriate choice in specific contexts: highly regulated content domains, institutional governance requirements, and genuinely stable content. Knowing when the exception applies is a professional judgment call.

Reflection Questions

For individual journaling or team discussion.

1. Think of a recent project in which the ADDIE sequence created friction, where you found yourself needing to revisit an earlier phase, compress a phase, or skip steps to meet a timeline. What specifically caused the friction, and which of ADDIE's structural limitations does it reflect?
1. Where in your current work do you already run Investigation and Design activities simultaneously, even if the formal process treats them as sequential? What would it look like to make that practice explicit and intentional rather than informal?
2. Consider your most recent program evaluation. When did the data arrive relative to when it was most useful? What would it have taken to act on evaluation insights in near-real-time rather than in a post-implementation report?
3. Using the comparison table in this chapter, identify the one or two dimensions where the difference between ADDIE and INDIE would have the greatest impact on a current or recent project. What would INDIE's approach have changed about the process or the outcome?
4. Are there programs in your portfolio that genuinely benefit from ADDIE's structured phases, the exception cases described at the end of this chapter? How do you know the difference between a legitimate exception and an organizational habit of staying comfortable?

Part Two · Chapter 6

IN Phase

INvestigation-Design: The Fusion That Changes Everything

The meeting request came in on a Wednesday afternoon. The subject line read: "New hire onboarding, need something by Q3." The sender was a vice president of operations Ruth had never spoken with. The body of the email contained three sentences: the organization had doubled in size over the past eighteen months, new hires were taking too long to reach full productivity, and someone had told him that training could help.

Ruth was a senior instructional designer with nine years of experience. She had built onboarding programs before. She knew, with the confidence of someone who had made this mistake and learned from it, that the request in this email was not yet a design brief. It was a symptom description. The actual problem, what was causing the slow ramp-up, whether training was the right intervention, what specifically new hires needed to know and be able to do and by when, was still entirely unknown.

She accepted the meeting. And then, before it happened, she did something that would once have been impossible in that window of time: she asked an AI tool to analyze the organization's publicly available information, the role profiles for the positions affected, and the three previous onboarding survey summaries she had found in the shared drive, and to produce a preliminary gap hypothesis and a set of questions she should be asking. When the draft arrived twenty minutes later, she revised it based on her own knowledge of the organization, added two questions the AI had missed, and removed one it had included that she knew would put the VP on the defensive.

She walked into the meeting not with a blank intake form, but with a working hypothesis and a set of well-targeted questions. That is what the IN phase looks like in practice. It is not the absence of analysis. It is analysis running alongside design from the first conversation.

The Logic of Fusion

The central premise of the IN phase, that investigation and design are not sequential activities but concurrent streams, is not simply a response to AI's capabilities. It is a recognition of something that experienced designers have always known, even when the formal model they worked within required them to pretend otherwise.

Investigation without design thinking produces data that nobody knows what to do with. The questions worth asking in a needs assessment depend on what kinds of solutions are possible, and knowing what solutions are possible requires design thinking. Conversely, design without ongoing investigation produces solutions to problems that may have been misunderstood from the start, or that evolved during a development cycle that moved too slowly to notice.

What AI has changed is not the underlying logic. It is the practical barrier to running these streams concurrently. When generating a preliminary curriculum draft required two days of work, it made sense to delay it until the analysis was substantially complete. When it requires forty minutes and a good prompt, the delay produces no quality benefit and considerable speed cost. The fusion was always theoretically sound. AI has made it operationally feasible.

Theoretical Grounding: Knowles and Vygotsky in the IN Phase

Malcolm Knowles' andragogy establishes that adult learners are problem-centered rather than subject-centered, are motivated by the immediacy of application, and bring existing experience that must be honored rather than ignored. These principles shape Investigation-Design in fundamental ways: the questions we ask must connect to real performance problems (not training topics), the objectives we write must reflect genuine job requirements (not content outlines), and the analysis must surface what learners already know (not just what they need to learn). Vygotsky's Zone of Proximal Development, the productive space between what a learner can do independently and what they can do with support, is the primary design target for the curriculum architecture produced in this phase. Finding the ZPD requires knowing both the current state (investigation) and the target state (design) simultaneously. That is the fusion.

The Five Streams of Investigation-Design

INDIE's IN phase is organized around five concurrent streams. They are not sequential steps. In practice, a designer moves between them continuously, with information discovered in one stream reshaping the work happening in the others. What follows is a description of each stream, the AI applications that accelerate it, and the human judgment that remains essential throughout.

Stream 1: Stakeholder Engagement: *Understanding context before content*

Every design project exists within an organizational context that shapes what is possible, what is necessary, and what will actually happen when the learning solution meets reality. Stakeholder engagement is the stream that surfaces that context: the business objectives driving the training need, the political dynamics that will affect adoption, the resource constraints that will bound what is buildable, and the definitions of success that will determine whether the program is considered effective.

This is also the stream most resistant to AI augmentation, not because AI cannot help with it, but because the most valuable work in stakeholder engagement is human: the conversation where a department head says something revealing she didn't intend to, the question that surfaces the real problem underneath the stated problem, the relationship-building that makes a stakeholder willing to be honest about constraints they would rather not name. AI can prepare you for those conversations and help you synthesize what you learn from them. It cannot have them for you.

AI Applications in Stream 1

- **Pre-interview preparation:** AI can analyze available organizational information, public documents, existing performance data, prior training records, and generate a preliminary gap hypothesis and targeted interview questions before the first stakeholder conversation. This allows designers to walk into initial meetings with a working hypothesis rather than a blank intake form, which changes the quality of the conversation.

- **Interview transcript analysis:** After stakeholder conversations, AI can process transcripts or detailed notes to identify recurring themes, surfaced tensions, stated versus implied priorities, and gaps in the information collected. A synthesis that might take half a day manually arrives in minutes, and the designer's energy goes to evaluating and extending the analysis rather than producing it.
- **Stakeholder communication drafting:** Meeting summaries, follow-up questions, project scoping documents, and expectation-setting communications can all be drafted with AI assistance. The designer's judgment about tone, political sensitivity, and relationship management remains essential; the drafting labor is substantially reduced.

PROMPT: Pre-Interview Preparation
I am preparing to meet with a senior stakeholder about a potential training initiative. Here is what I know so far: Organization/department: [DESCRIBE] Stated problem: [DESCRIBE WHAT WAS COMMUNICATED] Available background: [PASTE ANY RELEVANT DOCS OR SUMMARIES] Stakeholder role: [TITLE AND APPROXIMATE SCOPE] Known constraints: [TIME, BUDGET, AUDIENCE SIZE, ETC.] Please: 1. Generate a preliminary hypothesis about the actual performance gap (distinguish between training-addressable and non-training causes) 2. Identify the 5-7 most important questions to ask in this meeting 3. Flag any topics I should approach carefully given the organizational context 4. Note what information is missing that would significantly change the analysis Frame the questions using Knowles' adult learning principles — focus on performance problems, not training topics.

PROMPT: Stakeholder Interview Synthesis
I have completed [NUMBER] stakeholder interviews about [TRAINING TOPIC/NEED]. Here are my notes/transcripts: [PASTE NOTES OR TRANSCRIPTS] Please analyze these and provide: 1. The 4-6 most significant themes across all conversations 2. Points of stakeholder alignment (where perspectives agree) 3. Points of tension or disagreement (where perspectives conflict) 4. The gap between what stakeholders say they want and what the data suggests they actually need 5. Unanswered questions that require follow-up 6. Your assessment of whether a training intervention is the appropriate response to the identified problem, or whether other interventions should be considered Flag any statements that suggest the real problem may be different from the stated problem.

Stream 2 Needs Analysis: *Finding the gap that training can actually close*

Needs analysis is the systematic investigation of the difference between current and desired performance, and the determination of whether that gap is one that a learning intervention can close. The second half of that definition is as important as the first. Experienced performance consultants estimate that fewer than half of requests for training reflect problems that are primarily addressable through learning. The rest reflect feedback deficits, resource limitations, unclear expectations, motivational issues, or organizational systems failures that training cannot fix.

The IN phase takes this seriously. Investigation is not just about identifying that a gap exists. It is about understanding its causes well enough to know whether instruction is the right tool for closing it. An instructional designer who builds an excellent training program to solve a problem that is actually caused by a broken feedback loop or an unclear promotion criteria has done their job well and failed their organization entirely.

AI Applications in Stream 2

- **Performance data analysis:** AI can process quantitative performance data, productivity metrics, quality scores, error rates, time-to-proficiency measures, to identify patterns that indicate where the gap is largest and what it looks like at a behavioral level. This kind of analysis, done manually, requires significant time and statistical literacy. AI produces a usable first-pass analysis quickly, which the designer then interrogates and extends with their contextual knowledge.
- **Root cause classification:** Given a description of the performance gap and available context, AI can generate a structured root cause analysis distinguishing between knowledge/skill deficits (training-addressable), environmental factors, motivational factors, and systemic factors. This is not a substitute for the designer's judgment about which causes are primary, but it surfaces the full landscape of possible causes faster than manual analysis.
- **Learning gap prediction:** In organizations with historical performance data, AI can identify patterns in how skills develop over time and predict where future gaps are likely to emerge, enabling proactive curriculum planning rather than purely reactive response.

> **Theoretical Grounding: Piaget's Schema Theory in Needs Analysis**
>
> Jean Piaget's schema theory provides a useful analytical lens for the needs analysis stream: understanding not just what learners don't know, but the structure of what they do know and how new learning will relate to it. Some performance gaps reflect missing foundational schemas; the learner lacks the conceptual infrastructure that new knowledge would need to attach to. These require a different instructional approach than gaps that reflect outdated schemas, where the learner has existing knowledge that must be actively restructured rather than simply extended. The distinction shapes both content sequencing and the choice of instructional strategies, and it is worth building into AI prompts explicitly.

PROMPT: Needs Analysis and Root Cause Classification
I am conducting a needs analysis for [JOB ROLE / DEPARTMENT] at [ORGANIZATION TYPE]. Here is the available performance data: Performance metrics: [INSERT DATA] Error or quality patterns: [INSERT DATA] Time-to-proficiency data (if available): [INSERT DATA] Stakeholder descriptions of the problem: [SUMMARIZE] Please provide: 1. A structured analysis of the performance gap (current vs. desired state) 2. A root cause classification across four categories: - Knowledge/skill deficits (training-addressable) - Environmental factors (tools, resources, workspace) - Motivational factors (incentives, feedback, consequences) - Systemic factors (processes, organizational structures, culture) 3. Using Piaget's schema theory, assess whether the gap reflects: - Missing foundational schemas (learners lack conceptual infrastructure) - Outdated schemas needing restructuring (learners have prior knowledge that conflicts with the new learning) - Extension of existing schemas (learners have the foundation; they need to build on it) 4. Your assessment of what proportion of the gap is training-addressable 5. Recommended training priorities on a 1-5 urgency scale with justification Flag where the available data is insufficient to reach a confident conclusion.

Stream 3 Learner Analysis: *Designing for real people, not abstract audiences*

Every design decision in INDIE is ultimately a decision about a specific group of learners in a specific context with specific prior knowledge, specific constraints on their time and attention, and specific motivations for engaging, or not engaging, with the learning. Abstract audience descriptions ("mid-level managers", "front-line staff", "new hires") are not sufficient to drive good design decisions. Learner analysis produces the specificity that separates instruction designed for real people from instruction designed for a placeholder.

The outputs of learner analysis are typically two: a set of detailed personas that represent the range of people who will engage with the learning, and a clear picture of the Zone of Proximal Development – the productive learning space between what learners can currently do independently and what they are capable of with appropriate support. The ZPD is not a single point; different learners will have different ZPDs, and the curriculum architecture must accommodate that range.

AI Applications in Stream 3

- **Persona generation:** Given sufficient data on demographics, role requirements, experience levels, self-reported learning preferences, and job challenges, AI can produce detailed learner personas that ground design decisions in human specificity. These drafts require validation against actual learner and stakeholder input before they drive design decisions, but they provide a structured starting point that accelerates the process significantly.
- **Prior knowledge mapping:** AI can analyze available data on current competency levels, assessment results, performance records, onboarding surveys, to produce a map of the learner population's existing knowledge and skill distribution, identifying both the floor and the ceiling that the curriculum architecture must accommodate.
- **Motivational and barrier analysis:** AI can synthesize data from learner surveys and focus group notes to produce a structured analysis of what motivates the target audience to engage with learning and what barriers they face, practical constraints that should inform modality choices, time allocations, and support structures.

PROMPT: Learner Persona Development
I need to develop learner personas for [TRAINING PROGRAM] targeting [JOB ROLE]. Here is the available data on this audience: Demographics: [INSERT: experience range, educational background, etc.] Current skill/competency levels: [INSERT ASSESSMENT OR OBSERVATION DATA] Job challenges and pain points: [INSERT FROM INTERVIEWS OR SURVEYS] Self-reported learning preferences: [INSERT IF AVAILABLE] Practical constraints: [INSERT: available time, location, technology access, etc.] (Continued...) Please create 3-4 distinct learner personas. For each persona, include: - Name, role, and experience level - Current knowledge and skills relevant to this training - Primary job challenges this training should address - Motivations for engaging with the learning (WIIFM) - Barriers to engagement and learning transfer - Preferred learning modalities and formats - Vygotsky's ZPD: what they can do independently vs. what they can do with support Design the personas to represent meaningfully different points in the audience distribution, not just demographic variation. Note: These personas require validation against actual learner input before they drive design decisions.

Stream 4 Objective Development: *Write the destination before mapping the route*

Learning objectives are the most consequential output of the IN phase. They are the bridge between what the investigation has revealed and what the development phase will build. They are also the primary criterion against which the Enhancement phase will ultimately measure effectiveness. Getting them right is not a quality-of-life issue for the design process. It is the design process.

INDIE uses Mager's three-part format as the standard for learning objective quality: every objective should specify the Performance (what the learner will be able to do), the Condition (the circumstances under which they will do it), and the Criterion (the standard against which performance will be measured). Objectives that lack any of these three elements are not complete, regardless of how well-intentioned they are.

Bloom's Taxonomy provides the second quality dimension: every objective should be written at the cognitive level that reflects what the job actually requires. A technician who needs to apply a diagnostic procedure needs an application-level objective, not a knowledge-level one. A manager who needs to evaluate competing options needs an evaluation-level objective, not a comprehension-level one. Misaligning the cognitive level of an objective with the cognitive demand of the job is one of the most common and most consequential errors in instructional design, and AI can help catch it systematically.

AI Applications in Stream 4

- **Objective generation:** AI produces Mager-format objective drafts quickly, organized by Bloom's level and aligned to the performance gaps and job requirements identified in the analysis streams. These drafts are starting points for expert review, not finished products.
- **Objective critique and revision:** AI can evaluate a set of existing objectives against Mager's three-part format, flag objectives at the wrong Bloom's level for the stated job requirements, and suggest specific revisions. This is particularly useful in review sessions with stakeholders who have written their own objectives.
- **Objective-to-assessment alignment:** For each objective drafted, AI can propose the assessment approach most appropriate to its Bloom's level, catching in the IN-phase misalignments that would otherwise only surface in development.

PROMPT: Learning Objective Development
I am writing learning objectives for [COURSE/PROGRAM] targeting [AUDIENCE]. Performance gap identified: [DESCRIBE FROM NEEDS ANALYSIS] Job requirements: [DESCRIBE WHAT LEARNERS MUST DO ON THE JOB] Current competency level: [FROM LEARNER ANALYSIS] Business goals this training serves: [DESCRIBE] Available learning time: [HOURS/DAYS] Please generate 5-8 learning objectives using Mager's three-part format: "Given [condition], the learner will [performance verb] to [criterion]." For each objective: - Specify the Bloom's Taxonomy level (Knowledge, Comprehension, Application, Analysis, Synthesis, Evaluation) - Confirm the cognitive level matches the job requirement - Propose the most appropriate assessment approach for this objective Organize objectives in a logical learning progression from foundational to complex. Flag any objective where the Bloom's level required by the job exceeds what can reasonably be achieved in the available learning time.

PROMPT: Objective Quality Review
Please review the following learning objectives for [COURSE TITLE]: [PASTE OBJECTIVES] For each objective, evaluate: 1. Mager compliance: Does it specify Performance, Condition, and Criterion? If not, what is missing? 2. Bloom's alignment: What cognitive level does this objective target? Is that level appropriate for the job requirement it serves? 3. Measurability: Can performance be observed and assessed against the criterion? 4. Adult learning alignment: Does this objective connect to a real performance problem, or does it reflect a content topic? Provide a revised version of any objective that fails these criteria, with a brief explanation of what was changed and why.

Stream 5 Curriculum Architecture: *Mapping the learning journey.*

With objectives established and learner personas validated, the fifth stream of the IN phase produces the curriculum architecture: the organizational structure of the learning experience, the sequencing of content and activities, the modality selections for each component, and the scaffolding design that moves learners progressively from supported to independent performance.

Curriculum architecture is where the IN phase most clearly merges investigation and design. Every architectural decision reflects what the investigation has revealed: the sequencing reflects the learner's ZPD, the modality choices reflect the audience's constraints and preferences, the scaffolding design reflects the complexity of the skill being developed. And the architecture itself surfaces new analytical questions: where is prerequisite knowledge assumed that should not be? Where is the cognitive load likely to exceed what the target audience can handle? Where is a modality choice driven by convention rather than evidence?

AI Applications in Stream 5

- **Learning path optimization:** AI can analyze the set of objectives produced in Stream 4 and propose a curriculum sequence that minimizes prerequisites, distributes cognitive load appropriately across the program, and builds toward the most complex objectives through a logical progression of foundational and intermediate learning.
- **Modality recommendation:** Given the objective type, audience characteristics, available delivery infrastructure, and time constraints, AI can recommend the most appropriate instructional modality for each curriculum component, flagging mismatches between what a modality can deliver and what the objective requires.
- **Scaffold design:** AI can generate a scaffolding plan that maps the supports to be provided at each stage of the learning progression, and the points at which those supports should be reduced as learner competence develops.

PROMPT: Curriculum Architecture Design
I need to design the curriculum architecture for [PROGRAM TITLE]. Validated learning objectives: [PASTE OBJECTIVES] Learner personas: [SUMMARIZE KEY CHARACTERISTICS] Available learning time: [TOTAL HOURS, FORMAT: SINGLE EVENT / BLENDED / ONGOING] Available modalities: [e.g., eLearning, ILT, Virtual ILT, Job Aid, Coaching] LMS and technology constraints: [DESCRIBE] Transfer environment: [DESCRIBE THE JOB CONTEXT WHERE SKILLS WILL BE APPLIED] Using Vygotsky's scaffolding principles, please design a curriculum architecture that includes: 1. Module/unit structure with title and primary objectives for each 2. Recommended sequencing with rationale (why this order, not another) 3. Recommended modality for each module with justification 4. Scaffolding plan: what support structures are provided at each stage, and at what point are they reduced 5. Assessment checkpoint placement 6. Estimated time allocation per module 7. Prerequisites: what must be true about learners before each module Flag any place where the cognitive load of the proposed sequence may exceed what the target audience can manage, and suggest a mitigation.

The Worked Example: From Email to Curriculum Draft

Abstract descriptions of the IN phase are useful up to a point. What follows is a worked example that traces a real design project, Ruth's onboarding program from the opening scene, from the initial stakeholder email through the delivery of a first curriculum draft. The example is compressed; in practice, the work described below unfolded over ten days. What it illustrates is the parallel-stream logic of the IN phase: how investigation and design work inform each other continuously rather than in sequence.

The Project: New Hire Onboarding for a Scaling Operations Team

Day 1–2 *Pre-meeting preparation*	Ruth reviews three previous onboarding survey summaries, the role profiles for the five most commonly hired positions, and the organization's publicly available values and competency framework. She runs a pre-interview preparation prompt and receives a preliminary gap hypothesis (new hires lack organizational context and decision-making authority maps, not technical skills) and seven targeted interview questions. She revises two questions and removes one. She also asks the AI to generate three possible curriculum architectures based on the hypothesis, not to use them, but to have design possibilities in mind when the analysis begins confirming or disconfirming the hypothesis.

Day 3 *Stakeholder meeting + immediate synthesis*	The VP confirms that technical onboarding is working; the issue is that new hires don't understand how decisions get made, who to go to for what, or how their role connects to organizational strategy. He also reveals that two senior managers have refused to participate in previous onboarding efforts because they felt the content was beneath their team's expertise. Ruth notes this as a political constraint that will affect the design. Within two hours of the meeting, she has run an interview synthesis prompt on her notes and received a structured theme analysis. She adds three observations the AI missed about the manager resistance issue, which the tool had logged as "participation concerns" without recognizing its political significance.

Day 4–5 *Learner analysis + persona development*	Ruth distributes a short pre-analysis survey to twelve recent new hires across four departments. She also conducts two focus groups, each thirty minutes, with two cohorts of people who completed onboarding in the past six months. She runs the focus group notes through a persona generation prompt and receives four persona drafts. She validates these against her own observations and makes significant revisions to two: the AI had missed that the technical staff personas had much higher prior organizational knowledge than their tenure suggested, because many had transferred from a related division. The ZPD for this subgroup was substantially further along than the survey data alone indicated.

Day 6 *Objective development*	With the gap clearly identified (organizational context and decision-making navigation, not technical skills), Ruth runs an objective development prompt. The AI generates eight objectives; she evaluates each against Mager's criteria and the Bloom's level required by the actual job context. She revises three objectives: one was written at

	knowledge level (recall an org chart) when the job required application level (navigate to the right person for a specific category of decision). She reduces the eight to six, combining two that addressed the same underlying competency. She runs the revised set through an objective quality review prompt and receives confirmation on the Mager compliance of five objectives, with one remaining flag: the criterion on objective four is not measurable in the assessment context she's designing for.

Day 7–8 *Curriculum architecture*	Ruth runs a curriculum architecture prompt with the six validated objectives, persona summaries, and a constraint set: ninety minutes total, blended format (thirty-minute async module plus sixty-minute facilitated session), must work for the manager-resistant departments. The AI produces a three-module architecture with a sequencing rationale and scaffolding plan. Ruth makes one structural change: she moves the decision-mapping activity to before the facilitated session rather than within it, because she knows the manager-resistant departments will engage more freely if the initial work is done individually, not in a room with their manager watching. This is a judgment call driven by her knowledge of the organizational dynamics, information the AI did not have and could not have inferred from the prompt.

Day 9–10 *First draft review with stakeholder*	Ruth presents the six objectives and curriculum architecture to the VP. She brings the learner personas and the root cause analysis as supporting documents. The VP proposes adding a module on organizational history and values; Ruth explains, with reference to the persona analysis, that this content is not where the gap is and recommends addressing it through a reference document rather than instructional time. He agrees. The curriculum architecture moves into Development with stakeholder alignment on objectives, structure, and success criteria. Total time from initial email to approved curriculum draft: ten days.

Sidebar: When Stakeholders Expect a Formal Analysis Phase

One of the most common practical challenges in adopting the IN phase is organizational: many stakeholders, and some designers' own managers, have been trained to expect a formal, completed analysis deliverable before any design work begins. They may interpret the concurrent streams of the IN phase as cutting corners or skipping analysis, when it is actually more rigorous in its integration of analysis and design than the sequential approach it replaces.

There are two effective responses to this, and they are not mutually exclusive.

The first is to produce the analysis deliverable anyway, as a document that represents the findings of the investigation stream, separate from the curriculum architecture that represents the design stream. This satisfies the governance requirement without forcing the work itself to be sequential. The document is produced from the same work; it is the sequence of production rather than the sequence of work that changes. AI makes this easy: the synthesis prompts used during investigation produce outputs that can be compiled into a formal analysis report with minimal additional effort.

The second is to change the conversation about what analysis is for. The purpose of front-end analysis is to produce design-informing insight, not to produce a document. When analysis and design are concurrent streams that inform each other continuously, the analysis is more design-informing, not less. Making this argument requires being able to show stakeholders what the IN phase produces and how it produced it, which is why documentation of the reasoning at each stream is worth building into the workflow from the start.

A practical script: "We conducted the needs analysis and the initial design work concurrently, which allowed the design decisions to be tested against analytical findings as we went rather than after the analysis was complete. Here is the analysis document, and here is how each of the design decisions connects to what it found." That is not a defense. It is a demonstration of more rigorous practice.

What the IN Phase Produces

At the conclusion of the IN phase, a design project should have the following in hand. Each item is a deliverable in its own right; each also feeds directly into the Development phase that follows.

1. **A validated performance gap statement.** A clear, evidence-based description of the difference between current and desired performance, the root causes that are training-addressable, and the organizational context in which the learning solution will operate.
2. **A learner persona set.** Three to five detailed personas representing the range of people who will engage with the learning, validated against actual learner and stakeholder input, with ZPD estimates for each.
3. **A validated objective set.** Five to eight Mager-compliant learning objectives at the appropriate Bloom's levels, reviewed for measurability and alignment with actual job requirements.
4. **A curriculum architecture document.** Module structure, sequencing with rationale, modality selections with justification, scaffolding plan, assessment checkpoint placement, and time allocations, all mapped to the validated objectives and learner personas.
5. **A stakeholder alignment record.** Documentation of the agreements reached with key stakeholders on objectives, scope, success criteria, and constraints, the foundation for managing scope and expectations throughout the rest of the project.

One deliverable that is not on this list is a training plan that lists topics and durations. Topic lists are a common output of needs assessment processes, and they are almost never what the design process actually needs. An instructional designer who knows what the performance gap is, who the learners are, what they will be able to do at the end, and how the curriculum will be structured has everything they need to move into development. A list of topics tells them none of these things.

The Judgment the IN Phase Requires

It would be a mistake to close this chapter without naming what AI does not do in the IN phase, because the risk of over-relying on AI output is highest here, precisely because the outputs are structurally compelling and the errors are not always visible on the surface.

AI does not know your organization. Every persona draft, objective set, and curriculum architecture produced by an AI tool in the IN phase is built on the context you provided and the patterns in its training data. It does not know that the two senior managers who refused to participate in previous onboarding efforts are doing so because they resent the implication that their teams need additional training. It does not know that the VP who requested the program is three months into his role and has not yet established credibility with those managers. It does not know that the last onboarding program was cancelled halfway through because of a reorganization that is still politically raw. Ruth knew all of these things, and they shaped every design decision in the IN phase. The AI's curriculum architecture was a useful starting point. Ruth's revision of it was the actual design work.

AI does not exercise performance consulting judgment. The determination that a request for training reflects a training-addressable problem, rather than a feedback deficit, an unclear expectations issue, or a management problem wearing a training costume, requires human judgment applied to organizational knowledge. AI can generate a root cause framework. It cannot tell you which root cause is actually primary in your organization, with your stakeholders, in this particular political moment.

AI does not validate its own outputs. Every persona, every objective, every architectural recommendation produced in the IN phase is a hypothesis. Validation requires exposure to actual learners, actual stakeholders, and actual organizational context. Building that validation in, not as an afterthought but as a designed step in each stream, is the designer's responsibility, and it is not a step that AI can perform. The IN phase is faster with AI, but it is better because of the designer using it.

Chapter Summary

- The IN Phase fuses investigation and design into concurrent streams because they are not genuinely sequential, they have always informed each other continuously, and AI has made it operationally feasible to run them that way.
- The theoretical foundation for the IN phase is Knowles' andragogy (adult learners are problem-centered and motivated by immediate application) and Vygotsky's Zone of Proximal Development (the design target for curriculum architecture is the productive space between current and achievable performance).
- The five streams of the IN phase, Stakeholder Engagement, Needs Analysis, Learner Analysis, Objective Development, and Curriculum Architecture, run concurrently, with information discovered in each stream continuously reshaping the work in the others.
- AI accelerates every stream: pre-interview preparation, interview synthesis, root cause classification, persona drafting, objective generation and review, and curriculum architecture design. In each case, the AI produces a structured starting point that requires expert review, validation, and revision.
- The worked example demonstrates the parallel-stream logic in practice: investigation and design informing each other continuously across ten days, with AI producing structural frameworks and human expertise producing the decisions those frameworks alone could not make.
- The IN phase produces five deliverables: a validated gap statement, a learner persona set, a validated objective set, a curriculum architecture document, and a stakeholder alignment record.
- What AI does not do in the IN phase: know the organization, exercise performance consulting judgment, or validate its own outputs. The judgment the IN phase requires is irreducibly human.

Reflection Questions

For individual journaling or team discussion.

1. Think about the last needs analysis you conducted. At what point did design thinking begin influencing your analytical questions? If you had named that influence explicitly and worked with it intentionally, what might have been different about the process or the output?
2. Of the five IN phase streams, which one do you feel most confident AI can accelerate in your current work context? Which one makes you most cautious about over-reliance on AI output, and why?
3. Examine the learning objectives for a program you currently have in development or have recently completed. Apply Mager's three-part test and

Bloom's level check to each one. How many pass both? What does the failure pattern tell you about where the design process needs more discipline?

4. Ruth made a curriculum architecture change, moving the decision-mapping activity, based on organizational knowledge the AI did not have. Think of a design decision you have made recently that depended on contextual knowledge AI would not have had access to, even with a well-constructed prompt. How would you have flagged that decision explicitly in a documented design process?
5. If a stakeholder in your organization asked to see the completed analysis document before approving curriculum design work, how would you respond? What would you show them, and how would you explain the IN phase in terms that honored their process expectations while representing the actual work accurately?

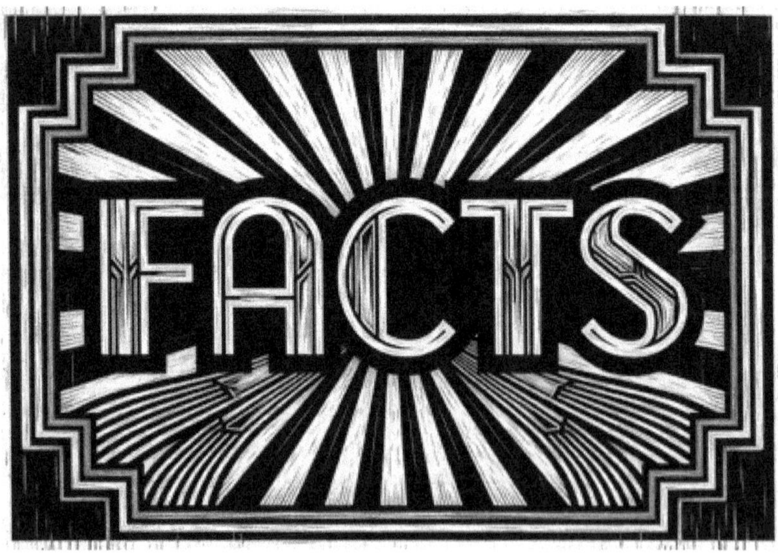

Part Two · Chapter 7

Phase D

Development: Building with Intelligence

Rafael had a problem that was not supposed to exist. He was an eLearning developer with seven years of experience in Articulate Storyline and a reputation for building modules that actually worked, that did what the storyboard said, delivered on time, and survived content review without collapsing into scope creep. He was fast, meticulous, and professionally secure in a way that came from knowing his craft. The problem was the storyboard in front of him.

It had been produced with AI assistance, a fact that was visible on every screen. The content was accurate. The scenarios were plausible. The objectives were clearly stated on slide three. But as Rafael read through it, a professional discomfort grew that he struggled to name precisely. The module opened with a definition. It moved through five content sections, each explaining a concept. It concluded with a ten-question knowledge check. It was, he finally decided, a very efficiently produced document that would produce a very efficiently produced training module that would leave learners with approximately the same competence they started with.

The problem was not that AI had been used to produce it. The problem was that the designer who used the AI had not given the tool anything to work with except content. No learning architecture. No sequence of instructional events. No framework for moving a learner from "I heard about this" to "I can do this." The AI had organized the content competently. Nobody had designed the learning.

Rafael sent the storyboard back with a single comment: "This is organized. It is not instructional. Can we talk about Gagné?"

Why Gagné is the Architecture

Robert Gagné's Nine Events of Instruction, first published in 1965 and refined through decades of subsequent research, remain the most practically useful framework available to instructional designers for a specific reason: they describe the internal conditions of learning rather than the external conditions of content delivery. They answer not "what should the module contain?" but "what must happen psychologically for learning to occur?"

The distinction is everything. A module organized around content topics delivers information in a logical sequence. A module organized around Gagné's events creates the cognitive and motivational conditions under which a learner can actually acquire, process, store, retrieve, and apply new knowledge or skill. The first is a well-organized presentation. The second is instruction.

INDIE's Development phase uses Gagné's Nine Events as its structural architecture for all instructional material creation. Every module, every course, every learning experience built in the Development phase should be auditable against these nine events, not as a checklist to be satisfied formally, but as a genuine test of whether the learning conditions have been established. A module that skips Event 3 (stimulate recall of prior learning) is not missing a box to tick. It is missing the cognitive bridge that allows new information to connect to what the learner already knows.

This chapter works through each of the nine events in detail: what it accomplishes psychologically, what it looks like in practice across different delivery formats, how AI accelerates its development, and where human judgment remains irreplaceable. It closes with a worked example tracing a single module from objective through QA-ready storyboard, and a Development QA checklist that synthesizes the chapter into a tool for ongoing use.

Theoretical Grounding: Cognitive Load Theory in Development

John Sweller's Cognitive Load Theory, developed alongside the Gagné tradition rather than from it, provides the essential complementary framework for the Development phase. CLT distinguishes between three types of cognitive load: intrinsic load (the inherent complexity of the material itself), extraneous load (load imposed by poor instructional design, unnecessary complexity in presentation, unclear navigation, irrelevant elaboration), and germane load (the productive cognitive effort that supports schema formation and learning). INDIE's development standard is simple to state and demanding to achieve: reduce extraneous load aggressively, manage intrinsic load through sequencing and scaffolding, and maximize germane load through well-designed practice and feedback. AI-generated content frequently fails this standard because it defaults to the extraneous: verbose explanations, redundant elaboration, decoration that competes with signal. The designer's job in the Development phase includes a constant cognitive load audit on every AI output before it moves forward.

The Nine Events: Theory, Practice, and AI

Each event section below follows the same structure: a description of what the event accomplishes and why it matters, what it looks like across delivery formats, how AI can accelerate development of this event, and a prompt ready for adaptation. The events are presented in sequence because that is how they are typically experienced by learners, but designers often develop them non-sequentially, and the prompts are designed to support that.

1	**Gain Attention** *Create the cognitive opening that makes everything else possible*

Attention is not courtesy. In instructional design, gaining attention is the act of interrupting the learner's existing cognitive state, whatever they were thinking about before they opened this module, and replacing it with a state of genuine curiosity or concern about the topic at hand. It is the cognitive opening through which all subsequent learning will enter.

The most common failure of Event 1 is substituting a statement for an experience. "Welcome to Module 3: Understanding Data Privacy Regulations." This tells the learner what they are about to do. It does not create any motivation to do it. Compare: a scenario in which a well-meaning employee shares a patient record with the wrong person, leading to a regulatory investigation that cost the organization $200,000 and the employee her job. Same topic. Different cognitive opening. The second version creates a state: concern, recognition, the slightly uncomfortable feeling of "this could happen to me," that prepares the learner to receive the instruction that follows.

Across delivery formats, Event 1 takes different shapes: a video scenario for eLearning, a provocative question or case study for ILT, a compelling statistic paired with a personal reflection prompt for blended programs. The format changes. The requirement does not: create a genuine cognitive state of readiness, not simply an announcement of what is about to happen.

AI Applications

- **Scenario generation:** AI produces multiple opening scenario drafts quickly, allowing designers to compare approaches and select the one with the strongest emotional and cognitive hook for this specific audience. The drafts almost always require revision for organizational authenticity, but generating five options in minutes rather than drafting one from scratch is a genuine acceleration.
- **Format adaptation:** Given a scenario developed for one modality, AI can adapt it for others, converting a video scenario script into a facilitator prompt and discussion question, or a text-based case into a branching scenario structure.

PROMPT: Event 1: Gain Attention
Create 3 distinct opening scenarios for a module on [TOPIC] for [AUDIENCE]. The scenarios should: - Connect directly to a real consequence of poor performance in this area - Be emotionally plausible for the [JOB ROLE] context - Create cognitive dissonance or concern that motivates engagement - Be completable in 2-3 minutes (eLearning) or 5 minutes (ILT) - Lead naturally into the statement of learning objectives For each scenario, provide: - Setting and character(s) - The situation and the decision point - The consequence that creates the hook - The transition question that bridges to the objectives Note: Specify the industry/organizational context: [DESCRIBE]. Avoid scenarios that place blame on individual learners; focus on systemic or situational factors that any reasonable person might encounter.

2 Inform Learners of Objectives

Make the destination visible before the journey begins

The presentation of learning objectives is one of the most consistently mishandled events in instructional design practice. It is either skipped entirely (learners never know what they are expected to be able to do), or it is handled in a way that treats the objectives as administrative disclosure rather than motivational framing.

Gagné's Event 2 serves two distinct purposes, and effective design must serve both. The first is cognitive: presenting objectives activates the learner's schema for the topic and helps them organize incoming information around a meaningful structure. Research on advance organizers, Ausubel's contribution to this conversation, confirms that learners who know what they are trying to learn retain information more effectively than those who encounter it without a framework. The second purpose is motivational: stated in the right language, objectives connect the learning to outcomes the learner actually cares about, answering the question they are almost certainly asking but rarely saying aloud: "What is this going to do for me?"

The practical standard: objectives should be presented in learner-facing language (not the Mager format used for design purposes), connected explicitly to job performance, and accompanied by a brief framing of why this matters for this specific audience. "At the end of this module, you will be able to identify a data privacy incident and initiate the correct reporting procedure within the required timeframe" is both cognitively and motivationally more effective than "Learning Objective 2: Compliance Reporting."

AI Applications

- **Objective translation:** AI can convert Mager-format design objectives into learner-facing language, connecting performance statements to job context and WIIFM framing. This is a small but time-consuming task that AI handles well.
- **Format variation:** Interactive objective presentations, a brief self-assessment, a sorting activity that reveals the objectives through learner input, a video in which a character from the opening scenario frames the objectives, can be drafted by AI and refined by the designer for technical feasibility.

PROMPT: Event 2: Inform Learners of Objectives
I have the following Mager-format learning objectives for a module on [TOPIC]: [PASTE OBJECTIVES] Please create a learner-facing presentation of these objectives that: - Translates each objective into plain language a [JOB ROLE] would recognize - Connects each objective explicitly to a job performance outcome they care about - Answers the implicit question: 'Why does this matter to me, specifically?' - Takes no more than 3 minutes to read or view - Is appropriate for [DELIVERY FORMAT: eLearning / ILT / blended] Additionally, propose one interactive alternative to a static objective slide: a brief activity that reveals or reinforces the objectives through learner engagement rather than passive reading.

3	**Stimulate Recall of Prior Learning** *Build the bridge between what they know and what they're about to learn*

This is the event most commonly skipped in AI-generated content, and its absence is detectable in the quality of learning that results. When new information is presented without connecting it to existing knowledge structures, it arrives in working memory without an anchor. It may be understood in the moment. It is unlikely to be retained or transferred.

Piaget's schema theory is the theoretical foundation here: new information is learned by being assimilated into existing schemas (where it fits what the learner already knows) or by triggering accommodation (where existing schemas are restructured to incorporate information that doesn't fit). Both processes require that existing schemas be active, available in working memory, accessible for connection. Event 3 activates them.

Knowles adds a dimension that Piaget does not: adult learners bring experience that is not merely cognitive infrastructure but a source of identity and professional self-concept. Recall activities that honor that experience, that begin from what learners already know and build forward, create engagement that activities beginning from a deficit framing do not. "What do you already know about data privacy?" is more likely to activate genuine cognitive engagement than "Many employees don't understand data privacy."

AI Applications

- **Recall activity design:** AI can generate multiple recall activity format, reflection prompts, diagnostic questions, sorting activities, experience-sharing prompts for ILT, for a given topic and audience. These drafts consistently require revision for tone (the "deficit framing" problem is common in AI output) and for alignment with what the learner analysis revealed about prior knowledge levels.
- **Misconception surfacing:** AI can generate a set of common misconceptions for a given topic area, which can be built into recall

activities designed to surface and address pre-existing incorrect schemas before new content is presented.

PROMPT: Event 3: Stimulate Recall of Prior Learning
I am designing Event 3 for a module on [TOPIC] for [AUDIENCE]. From my learner analysis, I know: - Current knowledge level: [DESCRIBE — e.g., foundational awareness, some prior training] - Likely correct prior knowledge: [DESCRIBE] - Likely misconceptions or outdated schemas: [DESCRIBE IF KNOWN] - Delivery format: [eLearning / ILT / blended] Please design 2-3 recall activity options that: - Begin from what learners already know, not from what they don't - Surface any common misconceptions that new content will need to address - Connect prior experience to the upcoming content explicitly - Are appropriate for the delivery format - Take no more than 5 minutes For each option, note whether it is more appropriate for novice or experienced audiences, and flag any risk of the activity feeling condescending to practitioners who have significant prior experience.

4	**Present Content** *Deliver information at the cognitive pace of learning, not the pace of coverage*

Event 4 is where the most content development effort is spent, and where the cognitive load trap is most likely to close on a designer who is producing with AI. The trap is this: AI is very good at generating comprehensive, well-organized content. Comprehensive and well-organized are not the same as learnable.

Miller's Law (the working memory capacity of approximately seven items, plus or minus two) has been refined by subsequent research, but its core implication survives: working memory is finite, information that exceeds its capacity does not get processed, and the presentation of more content than working memory can handle at one time produces the subjective experience of learning while delivering very little of it. AI generates content efficiently. It does not self-regulate for cognitive load.

INDIE's standard for Event 4: present a maximum of three to five key concepts per instructional unit. Each concept should be introduced, illustrated with a concrete example from the learner's actual work context, and then paused, creating space for working memory to process before the next concept is introduced. Vygotsky's scaffolding principle applies here directly: sequence from the familiar to the unfamiliar, from the concrete to the abstract, never the reverse.

AI Applications

- **Content chunking and outlining:** AI can take a body of content and produce a chunked outline organized around three to five key concepts, with supporting detail mapped to each. The designer's review should check that the chunk boundaries are cognitive rather than topical, organized around what working memory can hold, not around how a textbook would organize the subject.
- **Example generation:** For each key concept, AI can generate multiple concrete examples from different industry contexts, at different complexity levels, for different learner profiles. This is one of the highest-value AI applications in development: good examples are time-consuming to write and AI produces them quickly, though they consistently require revision for organizational authenticity and specificity.
- **Plain language conversion:** AI can convert SME-authored technical content into plain language appropriate for a specified audience, one of the most common and most tedious tasks in content development, handled well by current AI tools.

PROMPT: Event 4: Present Content
I need to develop the content presentation for a module on [TOPIC]. Learning objective this content serves: [PASTE OBJECTIVE] Target audience: [DESCRIBE — role, experience level, context] Available time for this section: [MINUTES] Source material: [PASTE OR DESCRIBE SOURCE CONTENT]

Please produce a chunked content outline that:

1. Identifies 3-5 key concepts (no more, prioritize ruthlessly)
2. For each concept, provides:
   - A plain-language explanation appropriate for this audience
   - One concrete example from the [INDUSTRY/ROLE] context
   - One common error or misconception to address
   - A cognitive pause indicator [where the learner should stop and process before moving on]
3. Sequences concepts from familiar-to-unfamiliar and concrete-to-abstract
4. Flags any concept where the intrinsic complexity is high enough to warrant additional scaffolding or prerequisite review

Apply Sweller's cognitive load principles: eliminate elaboration that does not serve the learning objective, and flag any section where extraneous load risk is high.

5 Provide Learning Guidance

Give learners the tools to make sense of what they've encountered

Learning guidance is the instructional infrastructure that helps learners process, organize, and encode content they have just been presented. It is distinct from the content itself: where Event 4 presents the information, Event 5 provides the structures that help a learner make sense of it, connect it to what they already know, and begin the process of encoding it into long-term memory.

Guidance takes many forms: worked examples that show the cognitive process of applying a concept (not just the outcome), mnemonics that provide retrieval cues for complex information, decision trees that scaffold procedural reasoning, comparisons that illuminate a new concept by contrasting it with a familiar one, expert tips that compress years of experience into actionable heuristics. What all of these share is that they do not simply repeat the content; they provide a different angle of approach that supports the cognitive work of learning.

The failure mode to avoid: treating Event 5 as a summary slide. A summary of content already presented adds little cognitive value. A worked example that shows a practitioner reasoning through a realistic application, narrating the decision points, adds significant value. The test: does this guidance help learners do something with the content, or does it simply restate the content in condensed form?

AI Applications

- **Worked example development:** AI can draft worked examples that narrate the application of a concept or procedure, including the expert reasoning that a novice would not otherwise see. These drafts are typically accurate in structure but thin on the organizational and contextual specificity that makes a worked example feel authentic to the learner.
- **Job aid and reference tool creation:** Decision trees, quick reference guides, procedural checklists, and comparison matrices are all forms of learning guidance that extend beyond the training event into the performance environment. AI produces solid drafts of all of these, and the structural logic is usually sound.

PROMPT: Event 5: Provide Learning Guidance
I need to develop learning guidance for [TOPIC] following content presentation. Key concepts just presented: [LIST THE 3-5 CONCEPTS FROM EVENT 4] Target audience: [DESCRIBE] Most common application context: [WHERE WILL LEARNERS USE THIS?] Known points of confusion or difficulty: [DESCRIBE IF KNOWN] Please develop the following guidance tools: 1. One worked example showing an expert applying [SPECIFIC CONCEPT] in a realistic [JOB ROLE] situation, with narrated decision points that make the expert reasoning visible 2. One job aid (decision tree OR quick reference OR checklist) that a learner could use during actual job performance, not just during training 3. Three expert tips that compress practitioner wisdom about common pitfalls, edge cases, or efficiency shortcuts in this domain For the worked example, show the reasoning process, not just the outcome.

For the job aid, optimize for use under realistic time pressure.

6	**Elicit Performance (Practice)** *Learning that doesn't require doing is not learning*

Event 6 is where the distinction between instruction and information delivery becomes undeniable. Information can be received passively. Learning, defined as a change in capability that is durable and transferable, requires the learner to actually do something with it. The cognitive science here is settled: retrieval practice (the act of recalling and applying information) is a more powerful driver of long-term retention than re-exposure to the information itself. We remember what we have used. We forget what we have only received.

INDIE's standard for practice design is that practice activities should require the cognitive operation that the objective requires. If the objective is at the application level of Bloom's taxonomy, the practice activity should require application, not recognition of the correct answer in a multiple-choice item, which is a recall task. If the objective is at the evaluation level, the practice should require the learner to make and defend a judgment, not to identify the judgment that an expert has already made.

The second standard: practice should be graduated. Three levels of complexity: foundational, applied, and integrated, allow learners at different points in the ZPD to engage productively. A single practice activity at one difficulty level serves the narrowest band of the learner population. Graduated practice, with appropriate scaffolding that is reduced as competence develops, serves the full range.

AI Applications

- **Scenario-based practice design:** AI excels at generating realistic workplace scenarios that serve as practice contexts. The scenarios are often well-structured but frequently require revision for organizational specificity, demographics, and the balance between realistic complexity and instructional clarity.

- **Distractor development:** For scenario-based and multiple-choice practice items, AI can generate plausible distractors, incorrect options that represent genuine misconceptions or common errors rather than obviously wrong answers. This is genuinely difficult to do well manually and AI handles it adequately, though the distractors always require expert review for accuracy and plausibility.
- **Graduated complexity design:** Given a single practice scenario, AI can produce versions at three complexity levels by adjusting the ambiguity of the situation, the amount of context provided, and the number of variables the learner must manage simultaneously.

PROMPT: Event 6: Elicit Performance (Practice)
I need to design practice activities for [SKILL/TOPIC] in a module for [AUDIENCE]. Learning objective: [PASTE OBJECTIVE — include Bloom's level] Job context where skill will be applied: [DESCRIBE] Delivery format: [eLearning / ILT / blended] Known audience characteristics: [RELEVANT PERSONA DETAILS] Please design 3 practice activities at graduated complexity levels: Level 1 (Foundational): High scaffolding, clear context, limited variables Level 2 (Applied): Moderate scaffolding, realistic ambiguity, job-authentic context Level 3 (Integrated): Minimal scaffolding, competing priorities, requires judgment For each activity: - Scenario setup (setting, characters, situation) - The task the learner must complete - The cognitive operation required (must match Bloom's level of the objective) - Success criteria (what does correct performance look like?) - Time allocation For Level 3: include at least one plausible distractor or red herring that requires the learner to exercise judgment rather than apply a procedure.

7	**Provide Feedback** *Close the loop between performance and understanding*

Feedback is the most consequential variable in the practice experience, and it is the one most frequently reduced to "Correct! Great job!" or "Incorrect. Please try again." Neither of these is instructional feedback. Both are evaluative responses that tell the learner whether they got it right without telling them anything about why, what the correct reasoning was, or what to do differently.

Effective feedback serves three cognitive functions: it confirms correct understanding and reinforces the schema that produced the correct response, it corrects incorrect understanding by explaining the error in terms of reasoning rather than just outcome, and it extends understanding by revealing something the learner did not know they did not know. A learner who gets an answer right for the wrong reason and receives only "Correct!" has not had their misconception addressed. A learner who gets an answer wrong because they applied the right principle to the wrong situation and receives only "Incorrect" has not had their confusion resolved.

Knowles' andragogy adds a dimension specific to adult learners: feedback should be framed in terms of performance improvement, not evaluation. Adults in professional contexts receive feedback through the lens of their professional self-concept. Feedback that feels like criticism of their competence rather than guidance toward better performance is likely to trigger defensiveness rather than reflection. The framing matters as much as the content.

AI Applications

- **Feedback script development:** For each practice activity, AI can develop feedback scripts for correct, incorrect, and partially correct responses, including the explanatory content that makes feedback instructional rather than merely evaluative. These scripts consistently require revision for tone and for the organizational specificity that makes examples feel authentic.

- **Error analysis:** AI can generate a taxonomy of common errors for a given skill or concept, which informs both distractor design in Event 6 and targeted corrective feedback in Event 7. This is a high-value application that significantly reduces the time required to produce informed feedback that addresses real misconceptions rather than generic errors.

PROMPT: Event 7: Provide Feedback
I need to develop feedback scripts for the following practice activity: Practice scenario: [DESCRIBE THE SCENARIO FROM EVENT 6] Correct response: [DESCRIBE WHAT CORRECT PERFORMANCE LOOKS LIKE] Common incorrect responses: [LIST IF KNOWN, OR ASK AI TO GENERATE] Audience: [DESCRIBE] Please develop feedback scripts for: 1. Correct response feedback: - Confirm the correct reasoning (not just the correct answer) - Reinforce the principle or rule that was applied - Extend understanding with one insight the learner may not have considered 2. For each plausible incorrect response: - Acknowledge what was reasonable about the incorrect reasoning - Explain specifically what was wrong and why - Redirect to the correct principle without simply stating the answer - Frame the correction in terms of performance improvement, not evaluation of the learner 3. A self-assessment prompt the learner can use after receiving feedback to consolidate the learning Tone: Collegial, non-condescending. Assumes professional adults who are capable of learning from mistakes without being made to feel diminished by them.

8	**Assess Performance** *Measure what actually matters, not what is easy to measure*

Assessment in INDIE is not a synonym for testing. It is the systematic collection of evidence that a learner has achieved the performance specified in the learning objectives. The distinction matters because it changes the question: not "how many correct answers did the learner get?" but "is there evidence that this learner can do what the objective required, to the standard the objective specified, under conditions the objective described?"

Mager's three-part objective format, Performance, Condition, Criterion, is the direct template for assessment design. Every assessment item should be traceable to a specific objective, should require the performance the objective specifies (not a proxy for it), should be conducted under conditions that match the objective's conditions, and should be scored against the objective's criterion. An objective that cannot be assessed in these terms is not a well-written objective. An assessment that does not trace to specific objectives is not a valid measure of learning.

The cognitive load caution applies here too: assessment design should sample the domain of performance specified by the objectives, not attempt to exhaustively test every piece of information presented in the module. Over-long assessments increase extraneous load, reduce completion rates, and often measure recall of details rather than demonstration of the competencies that actually matter.

AI Applications

- **Assessment item generation:** AI produces assessment items across multiple formats, scenario-based questions, case analysis items, performance checklists, open-ended reflection prompts, efficiently. Every generated item requires expert review for Bloom's level alignment, distractor quality, and objective traceability.

- **Rubric development:** For performance-based and open-ended assessments, AI can draft rubrics that specify the criteria for different levels of performance. These drafts are typically sound in structure but require calibration against actual examples of learner performance to be reliable.

PROMPT: Event 8: Assess Performance
I need to develop summative assessment items for [MODULE TITLE]. Validated learning objectives: [PASTE ALL OBJECTIVES FOR THIS MODULE] Target audience: [DESCRIBE] Assessment format constraints: [e.g., eLearning quiz, observed performance, written submission] Available assessment time: [MINUTES] Please develop a balanced assessment that: 1. Includes at least one item per learning objective 2. Uses the following item types [SELECT AS APPROPRIATE]: - Scenario-based multiple choice (application level or higher) - Case analysis (analysis or evaluation level) - Performance checklist (for procedural objectives) - Reflection prompt (for objectives involving professional judgment) For each item: - Identify the objective it assesses - Confirm the Bloom's level matches the objective - Provide correct answer and rationale - For multiple choice: provide 3 plausible distractors with the reasoning error each distractor represents - For performance checklists: include observable behavioral indicators Flag any objective that cannot be adequately assessed in the available format or time, with a recommendation for an alternative assessment approach.

9	**Enhance Retention and Transfer** *Design for what happens after the module closes*

Event 9 is the event that most training programs treat as an afterthought and that is most consequential for actual organizational performance. The research on training transfer is consistent and sobering: without deliberate design for retention and transfer, a significant proportion of learning that occurs in training does not appear in job performance. The training worked. The transfer didn't.

The mechanisms of transfer failure are well understood. Forgetting curves (Ebbinghaus's research, replicated extensively since) show that information not retrieved decays rapidly, and a training event that presents information once without subsequent retrieval practice is designed for forgetting. Transfer-appropriate processing, the principle that learning transfers most effectively when the conditions of learning match the conditions of performance, explains why training conducted in abstract contexts produces less transfer than training conducted in conditions that closely simulate the job environment. Performance support that is unavailable at the moment of need in the job environment will not be used, regardless of how well it was designed.

INDIE addresses Event 9 through three design mechanisms: spaced practice (distributing retrieval opportunities across time rather than concentrating them in a single event), contextual transfer (designing practice and performance support for the job context, not the training context), and social reinforcement (creating opportunities for the learning community to sustain and extend learning through peer interaction after the formal training event). Vygotsky's social learning theory is directly applicable here: learning sustained in community is more durable than learning sustained in isolation.

AI Applications

- **Spaced practice design:** AI can produce a spaced retrieval schedule, a set of brief practice prompts, distributed across days or weeks following the training event, designed to reactivate learning at the optimal points in the forgetting curve. These can be delivered via LMS push notifications, email nudges, or integration with workflow tools.

- **Performance support development:** Job aids, quick reference guides, and embedded performance support tools designed for the job context rather than the training context. These differ from the guidance tools in Event 5 in their design target: Event 5 supports in-training processing; Event 9 performance support is designed for use under realistic job conditions.
- **Transfer activity design:** Application assignments, action planning tools, and manager-facing follow-up guides that extend the learning beyond the training event into the performance environment.

PROMPT: Event 9: Enhance Retention and Transfer
I need to design a retention and transfer strategy for [MODULE/PROGRAM TITLE]. Learning objectives: [PASTE OBJECTIVES] Job context: [DESCRIBE WHERE AND HOW SKILLS WILL BE APPLIED] Transfer environment: [DESCRIBE SUPPORTS AND BARRIERS TO TRANSFER IN THE WORK ENVIRONMENT] Available follow-up touchpoints: [e.g., LMS push, email, manager check-in, team meeting, 30/60/90-day follow-up] Manager involvement available: [YES / NO / LIMITED] Please develop a retention and transfer plan that includes: 1. Spaced retrieval schedule: - 3 brief retrieval practice prompts at Day 3, Day 14, Day 30 - Each prompt should take 2-3 minutes and require active recall, not passive re-reading 2. Performance support tool: - Designed for use during actual job performance (not just training) - Optimized for the realistic conditions of use [TIME PRESSURE, TECHNOLOGY ACCESS, ETC.] 3. Application assignment: - A structured task the learner completes in their actual job within 2 weeks of training - Includes a brief reflection component that connects the experience back to the learning objectives

4. If manager involvement is available: a manager follow-up guide with 3 coaching questions for a 10-minute post-training conversation

The Worked Example: A Single Module from Objective to QA

What follows traces one module from Raphael's onboarding program, specifically, Module 2: "How Decisions Get Made Here," through the full development cycle, showing how the nine events translate into a concrete instructional sequence and where AI was used, where it was revised, and where human judgment produced outcomes the AI could not.

Module 2: How Decisions Get Made Here

Learning objective: Given a decision that falls within their operational scope, the new hire will identify the appropriate decision-maker and initiate the correct consultation process without escalating unnecessarily, confirmed by a scenario-based assessment scored against the organizational decision matrix.

Bloom's level: Application | Format: 30-min async eLearning + 60-min facilitated session | Audience: New hires, all departments

Event 1: Gain Attention

AI generated three opening scenarios. Raphael selected the second, a new hire who escalates a routine operational decision to a VP because she doesn't know the decision falls within her manager's authority, causing a visible delay on a client deliverable and leaving the VP wondering why his direct reports aren't empowering their teams. She revised the scenario significantly: the AI version had the VP respond with visible frustration, which felt punitive. She rewrote his response as confused and mildly concerned, more realistic and less likely to trigger a defensive rather than a curious response from learners.

Events 2 and 3: Objectives and Recall

Objectives were presented through a brief interactive activity: learners were shown five decision scenarios and asked to estimate "who would you go to?" before any content was presented. This served both events simultaneously, surfacing prior knowledge (most new hires had reasonable instincts about clear-cut cases and uncertainty about edge cases) while revealing the objectives through the experience of realizing where instinct wasn't sufficient. AI drafted the five scenarios; Raphael revised two for organizational authenticity and removed one that inadvertently made the correct answer obvious.

Event 4: Content Presentation

Three concepts: how the organizational decision matrix works, the difference between decisions requiring consultation and decisions requiring approval, and the practical signals that indicate a decision is approaching a scope boundary. AI produced a content outline organized around six topics; Raphael consolidated to three concepts and added the practical signals section, which the AI had omitted because it was not present in the source material she had provided. This was a significant gap that would have produced content accurately describing the formal matrix without equipping learners to recognize when they were approaching its edges, which was the most common point of failure in the learner analysis.

Event 5: Learning Guidance

A worked example showing a new hire thinking through a decision in real time, narrating the signals she was looking for and the consultation she initiated. AI produced the structural draft; Raphael made it specific to the organization's actual decision categories and added a moment in which the character pauses and says "I'm not sure, let me check the matrix" rather than proceeding confidently. That moment of visible uncertainty followed by competent resolution was pedagogically important: it modeled the behavior the program was trying to build, which was not confident expertise but competent navigation of uncertainty. AI had not included it.

Events 6 and 7: Practice and Feedback

Three graduated scenarios. Level 1: a clear-cut case where the decision category is obvious and the matrix gives an unambiguous answer. Level 2: a case where the decision sits at a boundary between two categories, requiring the learner to recognize the ambiguity and initiate consultation. Level 3: a case with time pressure, a manager who is unavailable, and a client who is asking for an immediate answer, requiring the learner to make a judgment about when to proceed and when to escalate. AI drafted all three; Raphael revised Level 3 substantially. The AI version resolved the tension too cleanly. The real world, she noted in the design document, does not always offer clean resolutions, and a module that only shows correct application is not preparing learners for the actual complexity of the job.

Events 8 and 9: Assessment and Transfer

Assessment: a five-item scenario-based quiz, each item tracing to the module objective, with distractors representing the actual errors identified in the needs analysis. AI produced all five items; one was removed because the scenario inadvertently signaled the correct answer through the emotional valence of the character's response. Transfer: a structured observation task in the first two weeks, learners were asked to note three real decisions they encountered and identify where each fell in the decision matrix. A brief structured reflection connected the observation back to the module objective. This was Raphael's design, developed without AI assistance, because it required knowledge of the learners' actual work environment and a judgment about what would be practically feasible given their workload.

The Development QA Checklist

Quality assurance in INDIE is not a gate that content passes through at the end of development. It is a running standard applied at every stage, a set of criteria that each element of the developing module is held against before it moves forward. The checklist below synthesizes the standards described across this chapter into a tool for ongoing use.

Criterion	INDIE Standard
Gagné coverage	All nine events are addressed. No event is present only in name, each serves its genuine instructional function.
Attention (Event 1)	Opens with a scenario or hook that creates genuine cognitive readiness, not an announcement of what is about to happen.
Objectives (Event 2)	Presented in learner-facing language connected to job performance. WIIFM is explicit.
Recall (Event 3)	Activates prior knowledge explicitly. Begins from what learners know, not from what they lack. Surfaces likely misconceptions.

Content chunking	Maximum 3-5 key concepts per instructional unit. Cognitive pauses designed in. Sequenced from familiar to unfamiliar.
Cognitive load	Extraneous load minimized: no elaboration that does not serve the objective, no decorative elements competing with signal.
Examples	Each concept illustrated with at least one example authentic to the learner's job context. Examples are specific, not generic.
Guidance (Event 5)	Includes at least one worked example that narrates expert reasoning. Job aids designed for performance environment, not training environment.
Practice level	Practice activities require the cognitive operation specified by Bloom's level of the objective. Not just recall if objective is application or higher.
Practice graduation	Three complexity levels represented. Scaffolding present at Level 1, reduced at Level 2, removed at Level 3.
Feedback quality	Feedback is instructional, not merely evaluative. Explains correct reasoning, addresses incorrect reasoning, avoids condescension.
Assessment traceability	Every assessment item traces to a specific objective. Item format requires the performance the objective specifies.
Distractor quality	Multiple-choice distractors represent genuine misconceptions or common errors. Not obviously wrong.
Transfer design	Spaced retrieval plan in place. Performance support designed for job context. Application assignment specified.
Accessibility	All multimedia has captions or transcripts. Color is not the only carrier of meaning. Navigation is keyboard-accessible.
AI output review	All AI-generated content has been reviewed by a qualified designer for accuracy, organizational authenticity, cognitive load, and bias.

The Cognitive Load Trap: A Closer Look

The sidebar promised in the chapter outline is worth delivering at length, because the cognitive load trap is the most common and most consequential failure mode in AI-augmented development, and understanding it precisely is the prerequisite for avoiding it.

The trap works like this. A designer uses an AI tool to generate content for Event 4. The tool produces a thorough, well-organized explanation of the topic, comprehensive, accurate, clearly structured. The designer, under time pressure, accepts the output with minimal revision.

The resulting module presents more information per screen than working memory can process, uses elaboration that is accurate but not relevant to the learning objective, includes examples that are generic rather than contextually specific, and sequences concepts logically (in the sense that a textbook is logical) rather than pedagogically (in the sense that a learning experience is pedagogical). Learners complete the module. They score adequately on the knowledge check because the knowledge check tests recall of what was just presented. They return to the job and perform as they did before, because what they encountered was a comprehensive presentation of information, not instruction.

The trap is invisible in the development process because AI-generated content looks like good content. It is grammatically correct. It is well-organized. It is comprehensive. None of these are the right criteria. The right criteria are: does this create the conditions for learning? Does it respect the limits of working memory? Does it connect to what learners already know? Does it require them to do something with the information rather than receive it? Does it prepare them for the practice that follows?

Three practical habits protect against the cognitive load trap. First: always audit AI-generated content against the objective, not against the source material. Content that is accurate but not relevant to the objective is extraneous load. Remove it. Second: apply a strict concept count. If you count more than five distinct concepts in a single instructional unit, the unit needs to be divided or the content needs to be cut. Third: read every screen from the perspective of a learner encountering this information for the first time, with the specific prior knowledge described in the learner personas. If anything on that screen would be confusing, redundant, or irrelevant from that perspective, it does not belong there. Rafael's instinct, that the AI-generated storyboard was organized but not instructional, was a cognitive load diagnosis. He had seen the trap, and he knew what it produced. The Gagné framework was his tool for naming what was missing. That is the Development phase in practice: not just building content, but building the conditions under which learning can occur. Faster, now, with AI. Better, still, because of the designer.

Chapter Summary

- Gagné's Nine Events of Instruction is INDIE's structural architecture for Development because it describes the internal psychological conditions required for learning — not the external organization of content. A module organized around nine events creates instruction; a module organized around content topics creates a presentation.
- Cognitive Load Theory (Sweller) is the essential companion framework: Development must minimize extraneous load, manage intrinsic load through sequencing and scaffolding, and maximize the germane load that produces durable schema formation. AI-generated content frequently adds extraneous load through comprehensiveness that exceeds working memory capacity.
- AI accelerates development across all nine events: scenario generation, objective translation, recall activity design, content chunking, worked example drafting, practice scenario development, feedback scripting, assessment item generation, and spaced retrieval design. In each case, AI produces structured starting points that require expert review and revision.
- The cognitive load trap is the most common AI-augmented development failure: comprehensive, well-organized content that is not instructional because it has not been filtered against the objective, does not respect working memory limits, and does not create the cognitive conditions for learning.
- Three habits protect against the cognitive load trap: audit against the objective (not the source material), enforce a maximum of five concepts per instructional unit, and read every screen from the perspective of the learner persona.
- The Development QA checklist synthesizes the standards of the chapter into a running audit tool that should be applied at every stage of development, not only at the end.
- What AI does not do in the Development phase: recognize when content is pedagogically weak despite being accurate, catch the organizational and contextual specificity that makes examples authentic, or hold the quality standard that defines effective instruction. Those are the designer's job.

Reflection Questions

For individual journaling or team discussion.

1. Apply the Gagné audit to a module you have developed recently or are currently developing. Which events are genuinely present? Which are present in name only? Which are missing? What would it take to address the gaps?
2. Rafael's diagnosis of the AI-generated storyboard was that it was "organized but not instructional." Have you encountered content that fits this description in your own work or in work you have reviewed? What specifically made it organized without being instructional, and what would have changed it?
3. Of the nine events, which one do you consider most consistently underserved in the training programs you have encountered? What organizational or process pressures do you think produce that pattern?
4. The worked example showed Raphael making four specific revisions to AI-generated content that the tool could not have made on its own. Think of a development task you have recently completed with AI assistance. What revisions did you make that required knowledge or judgment the AI did not have? Were those revisions visible in the final product, or were they invisible to the people who reviewed it?

Part Two · Chapter 8

Phase I

Implementation: Agile Deployment for Real Organizations

The program was good. Sonia knew it was good because she had spent eight weeks making it good: the stakeholder interviews, the persona validation, the objective revision cycle, the content review with three subject matter experts who disagreed with each other on almost everything and somehow arrived at materials she was genuinely proud of. The facilitation guide was detailed without being prescriptive. The eLearning module was clean. The assessment was rigorous. She had done the work. Then, the launch happened.

Two of the five scheduled sessions were cancelled the morning they were supposed to run because a regional manager decided his team couldn't spare the time during quarter close. The facilitator for the third session had received the guide forty-eight hours before delivery and was visibly underprepared. The fourth session ran well with an experienced facilitator who knew the material and could improvise when the technology failed in the first twenty minutes. The fifth session was attended by six people, two of whom had come to the wrong room and stayed out of politeness.

Sonia sat with the evaluation data three weeks later and felt a particular kind of professional grief that designers who have been in the field long enough will recognize: the grief of good work that met a bad implementation. The program had not failed. The launch had failed. And in the data, and in the organizational memory, those two things were not distinguishable.

The lesson she drew from it was not that she needed to design better. It was that she needed to think about implementation the way she thought about design: systematically, anticipating failure points, building in redundancy, and treating the first delivery not as a launch but as a learning event.

The Launch Fallacy

The most consequential assumption embedded in traditional implementation thinking is that a program's first delivery is its definitive delivery, that the work of implementation is to execute the plan that development produced, and that the quality of implementation is measured by how faithfully that execution occurred. On this view, a cancelled session is a logistics failure. An underprepared facilitator is a coordination failure. A room that doesn't fill is a scheduling failure. The solution to each is better logistics, better coordination, better scheduling.

INDIE starts from a different premise: that the first delivery of any program is a source of information, not a definitive outcome. The information it generates, about what works with real learners in real organizational conditions, what the facilitation challenges actually are, what the audience characteristics the personas did not fully capture, what the transfer environment actually looks like, is more valuable than any amount of additional design review. And that information is only available through delivery.

This premise has a concrete implication: the goal of the first delivery is not to execute perfectly. It is to learn fast enough to improve before the full rollout. A pilot with twelve people that generates specific, actionable data on three design problems is worth more to the program's long-term quality than a flawless execution of a design that has never been tested with real learners.

The shift from "launch" thinking to "pilot" thinking is the central reorientation of INDIE's Implementation phase. It changes what designers plan for, what they measure, and what they consider success. It also changes the conversation with stakeholders, which requires its own skill and its own preparation.

Theoretical Grounding: Constructivism and the Implementation Environment

The social constructivist tradition, rooted in Vygotsky, extended by researchers like Lave and Wenger in their work on situated learning and communities of practice, holds that learning is not a private cognitive event but a social one, shaped by the context in which it occurs, the relationships through which it is mediated, and the community into which the learner is being inducted. This has direct implications for implementation: the organizational context of delivery is not just a logistical backdrop. It is part of the learning environment. A session delivered by a facilitator who is clearly knowledgeable and respected by the audience produces different learning than the same session delivered by a facilitator who is unknown to the audience or who is visibly uncertain about the material. The social signals of the delivery environment shape what learners believe about the content, how much cognitive effort they invest, and whether they carry the learning back to their work. Implementation design that ignores these signals will produce outcomes the design documents did not predict.

Designing the Pilot

A pilot is not a trial run of the full program at smaller scale. It is a deliberately designed learning event, for the design team, whose goal is to generate specific, actionable information about what needs to change before the full rollout. That distinction matters because it changes what a well-designed pilot looks like.

A well-designed pilot is not the most favorable possible first delivery, delivered by the most experienced facilitator, to the most enthusiastic cohort, under the most controlled conditions. That kind of pilot produces the least useful data, because its conditions are not representative of what the full rollout will encounter. A well-designed pilot is a representative first delivery: conditions close enough to the full rollout to reveal the problems the full rollout will face, controlled enough to allow observation and data collection, and small enough to allow rapid adjustment before those problems affect a larger population.

Who to Include in the Pilot

The pilot cohort should represent the range of the full audience, not its best-case segment. Include people who are skeptical of training as well as people who are enthusiastic about it. Include people whose roles create different relationships with the content. Include at least one person who is representative of the audience segment the personas suggested would find the program most challenging.

Size matters, but not in the direction most designers assume. A pilot cohort of eight to fifteen people is sufficient to surface the majority of significant design problems. A pilot cohort of thirty is not three times more useful; it is three times harder to observe and produces data that takes three times longer to process. Smaller is faster and, in a pilot context, faster serves quality.

One specific inclusion worth building in deliberately: include at least one person who is connected to the informal information network of the organization, someone whose response to the program will be shared with colleagues before the full rollout. Pilot participants do not experience the program in isolation. They talk about it. A pilot cohort that includes someone whose social network reaches the full rollout audience is a pilot cohort whose experience will pre-shape the audience's expectations. Design for that.

What to Observe

Pilot observation requires a deliberate plan, not passive attention. Before the pilot, identify the specific questions you most need answered: Which activities generated the most engagement? Where did confusion surface? How did the audience respond to the scenario in Event 1? Did the time allocations hold? Were there moments when the facilitation guide clearly wasn't sufficient? Design an observation protocol that directs attention to those questions rather than leaving the observer to decide in the moment what matters.

Assign at least one person to observe who is not facilitating. A facilitator who is simultaneously managing the room, monitoring time, and trying to observe learner responses is not observing — they are surviving. The observation data that matters most often comes from watching the learners rather than the facilitator: the moment of visible confusion during the scenario, the pair who whisper to each other during the practice activity, the person who answers the assessment confidently but incorrectly.

What Counts as a Successful Pilot

A successful pilot is one that generates specific, actionable information about what needs to change. By this standard, a pilot in which everything went smoothly and the feedback was uniformly positive has not been fully successful; it has generated evidence that the program works for this cohort under these conditions, which is useful but incomplete.

A pilot in which three specific design problems were identified, documented, and addressed before the full rollout is more successful, regardless of how rough the delivery felt.

This is a reframing that is worth making explicit with stakeholders before the pilot happens. If stakeholders are expecting a flawless delivery as the pilot's success criterion, they will interpret the identification of design problems as evidence of poor quality rather than evidence of a well-functioning improvement process. Setting expectations before the pilot is not spin. It is stakeholder alignment, and it is part of the designer's job.

> *A pilot in which three design problems were identified and fixed is more successful than one in which everything went smoothly. Problems found early are gifts.*

Sidebar: The One-Week Feedback Sprint

The most common reason pilot data doesn't produce program improvements is not that the data was poor. It is that the data sat in a spreadsheet for two weeks while everyone waited for a meeting that got rescheduled. The one-week feedback sprint is a structured process for moving from pilot delivery to design revisions in seven days.

Day 1 *Capture*	Within 24 hours of the pilot delivery, all observers and the facilitator complete a structured debrief template. Template covers: what worked well and why, what didn't work and the specific moment it became visible, what participants said in informal conversation (hallway feedback is often more honest than formal surveys), and what the observer would change if they could do one thing differently. AI synthesizes all debrief inputs into a consolidated themes document within the hour.

Day 2 *Analyze*	Participant reaction data (Level 1) arrives and is analyzed alongside the observer debrief. AI processes open-ended responses and identifies themes. The design team, no more than three people, meets for sixty minutes to categorize findings: what is a design problem (requires content or structural change), what is a delivery problem (requires facilitator guidance or support), and what is a context problem (requires stakeholder or logistics work). Do not confuse these. A

	design solution for a delivery problem changes the content. A delivery solution for a design problem adds facilitation burden that won't scale.

Day 3–4 *Prioritize & Revise*	Address design problems by category: high-impact and low-effort changes go first. AI drafts revised content sections, updated scenario language, additional facilitator guidance, or revised assessment items based on specific problem descriptions. Designer reviews and approves all revisions against the original learning objectives. A revision that fixes a delivery problem but creates a design problem is not a net improvement.

Day 5 *Validate*	Run revised materials through a fast internal review: one person who was in the pilot room, one who wasn't. The person who was in the room checks whether the revision addresses what they observed. The person who wasn't checks whether the revision introduces any new problems the pilot didn't surface. This is not a full design review. It is a thirty-minute sanity check with a narrow and specific question.

Day 6–7 *Communicate & Set*	Brief the program sponsor on what was found, what was changed, and why. This communication has two purposes: it maintains stakeholder confidence by demonstrating that the pilot process worked as designed, and it establishes a record of design decisions that will inform the Enhancement phase later. AI drafts the summary. The designer personalizes it for the specific audience and relationship. Full rollout materials are finalized.

(Continued...)

The sprint works because it compresses what would otherwise be a two-to-four-week improvement cycle into seven days. The compression is possible because AI handles the synthesis and drafting labor, because the problem categories are established on Day 2 before anyone writes a single word of revised content, and because the scope is constrained: the sprint addresses what the pilot revealed, not everything that anyone thinks could be improved.

AI Applications in the Implementation Phase

Implementation's AI applications are different in character from those of the IN and Development phases. Where earlier phases use AI primarily for analytical and generative work, Implementation uses it primarily for organizational and communicative work: managing the logistics of a deployment, synthesizing feedback data rapidly, supporting facilitators, and maintaining the stakeholder communications that keep a rollout on track.

Project Planning and Timeline Management

A program deployment involves more moving parts than the design phase typically anticipates: scheduling across multiple cohorts and locations, coordinating facilitator preparation, managing technical setup, communicating with participants and their managers, and tracking completion against a rollout timeline that is almost always tighter than originally planned. AI can generate deployment plans, identify dependency risks, draft coordinator communications, and produce tracking tools, tasks that collectively consume significant project management time that could otherwise go to the human work of relationship and quality management.

PROMPT: Deployment Planning
I need to plan the full rollout of [PROGRAM TITLE] following a successful pilot. Total audience size: [NUMBER] people across [NUMBER] locations/departments Available delivery windows: [DESCRIBE – days of week, time constraints] Facilitators available: [NUMBER AND EXPERIENCE LEVEL] Technical infrastructure: [LMS, video platform, room requirements] Rollout deadline: [DATE] Program format: [DESCRIBE – length, modality, cohort size] Please produce: 1. A phased rollout schedule organizing [NUMBER] cohorts across the available windows, with buffer time for facilitator preparation between sessions 2. A dependency map identifying which tasks must be complete before each rollout milestone can proceed (continued...) 3. A risk log covering the 5 most likely deployment risks with probability assessment and mitigation actions for each 4. A communications calendar: what to send, to whom, and when, from four weeks before first session through two weeks after final session Flag any constraint conflict that makes the current timeline unrealistic, with a specific recommendation for resolution.

Participant and Manager Communications

The communications that surround a program's delivery are not administrative overhead. They are part of the learning design. A participant who receives a thoughtfully framed pre-session message that explains why this program matters to their specific role, what to expect, and how to prepare arrives in a different cognitive and motivational state than one who receives a calendar invite with no context. A manager who receives a brief guide on how to support transfer after the session is more likely to have the thirty-second reinforcement conversation that doubles retention than one who has no idea what their team just experienced.

AI drafts all of these communications quickly. The designer's role is to ensure they reflect the actual audience, the actual program, and the actual organizational context, and to add the human specificity that makes communications feel like they came from a person rather than a template.

PROMPT: Pre-Session Participant Communication
I need to write a pre-session message for participants in [PROGRAM TITLE]. Audience: [JOB ROLE AND DEPARTMENT] Session format: [LENGTH, MODALITY, LOCATION/PLATFORM] Primary learning objectives (in plain language): [DESCRIBE] What participants should do to prepare: [DESCRIBE, OR 'none required'] What the program will and will not cover: [BRIEF SCOPE DESCRIPTION] Organizational context (why this program, why now): [DESCRIBE] Sender: [NAME AND ROLE — program sponsor, L&D, manager, etc.] Please draft a message that: - Takes 90 seconds or less to read - Answers the implicit question: 'Why should I care about this?' - Sets accurate expectations without over-promising - Is warm in tone without being performatively enthusiastic - Does not use L&D jargon (no 'learning journey', 'upskilling', etc.) Also draft a separate 3-sentence version for use as a calendar invite description.

PROMPT: Manager Transfer Support Guide
I need to create a brief manager guide for supporting transfer after [PROGRAM TITLE]. What participants learned: [SUMMARIZE KEY COMPETENCIES IN PLAIN LANGUAGE] How managers will see the learning appear in behavior: [DESCRIBE] The 2-3 most common barriers to transfer in this environment: [DESCRIBE] Time available for manager engagement: [REALISTIC ESTIMATE — 5-10 mins] Please produce a one-page manager guide that includes: 1. What your team member experienced (2-3 sentences, no jargon) 2. What you might notice differently in their work in the next 30 days 3. Three coaching questions for a 5-minute follow-up conversation (questions that open dialogue, not questions that test recall) 4. One thing to avoid: the most common manager response that undermines transfer in this specific context Tone: Collegial peer-to-peer, not top-down directive. Managers are partners in transfer, not compliance officers.

Facilitator Preparation and Support

Facilitation quality is the single highest-leverage variable in the implementation of instructor-led and blended programs. A well-designed program delivered by an underprepared facilitator produces worse outcomes than a less polished program delivered by someone who genuinely knows the material and the audience. This is not a design problem. It is an implementation problem, and it has a specific cause: facilitation preparation is under-resourced in most organizations, and facilitators are frequently given materials too late for meaningful preparation.

INDIE builds facilitator preparation into the implementation timeline as a non-negotiable milestone, not an optional step that gets compressed when the schedule is tight. The minimum preparation standard: facilitators should have materials at least one week before delivery, should have access to a preparation guide specific to this program (not generic facilitation advice), and should have the opportunity to ask questions of the designer before delivery.

AI accelerates the production of facilitator preparation materials, which have historically been under-developed because they sit at the end of the development timeline and absorb the time compression that the earlier phases have already experienced. With AI handling the drafting, preparation guides can be comprehensive without being a final-week emergency.

PROMPT: Facilitator Preparation Guide
I need to create a facilitator preparation guide for [PROGRAM/SESSION TITLE]. Session length: [DURATION] Participant profile: [DESCRIBE AUDIENCE] Facilitator background: [DESCRIBE — subject matter expert? ID professional? First-time facilitator for this content?] Delivery format: [IN-PERSON / VIRTUAL / HYBRID] Materials provided: [LIST WHAT THE FACILITATOR WILL HAVE] Please produce a preparation guide that includes: 1. The 5 things most important to know about this audience before walking in 2. The 3 moments in this session where facilitation skill matters most, and what skilled facilitation looks like at each moment 3. The 3 most likely facilitation challenges and a specific response to each 4. Questions participants are likely to ask, organized by section, with suggested responses 5. Technology and logistics checklist (items to confirm 24 hours before, 30 minutes before, and immediately before session start) 6. A suggested personal preparation routine for the hour before the session This guide is for the facilitator's use, not for participants. Write in a direct, collegial voice — peer practitioner to peer practitioner.

Rapid Feedback Analysis

The value of Implementation's feedback data depends entirely on how quickly it is processed and acted on. Level 1 reaction data collected after each session, synthesized manually, and compiled into a report three weeks after the final session is retrospective documentation. The same data synthesized within twenty-four hours of each delivery and used to adjust the next session is a design tool.

AI enables the speed that turns feedback into a real-time improvement mechanism. Open-ended survey responses, facilitator debrief notes, observer field notes, and informal participant comments can all be synthesized rapidly, producing a themes analysis, a priority ranking of issues to address, and a set of specific revision recommendations that the design team can act on before the next cohort arrives.

PROMPT: Rapid Feedback Synthesis
I need to synthesize pilot feedback for [PROGRAM TITLE] rapidly. Participant reaction survey responses: [PASTE QUANTITATIVE RATINGS AND OPEN-ENDED RESPONSES] Facilitator debrief notes: [PASTE OR SUMMARIZE] Observer field notes: [PASTE OR SUMMARIZE] Informal participant comments captured post-session: [PASTE IF AVAILABLE] Number of participants: [NUMBER] Please produce: 1. A themes analysis of the open-ended feedback (5-7 themes, frequency-weighted, with representative quotes for each) 2. A categorized issue list organized by: - Design issues (content, structure, activities, assessments) - Delivery issues (facilitation, pacing, technology) - Context issues (scheduling, environment, audience preparation) 3. A priority ranking of the top 5 issues to address before the next cohort, with a brief rationale for each ranking 4. Specific revision recommendations for each high-priority design issue Flag any finding that suggests the program objectives may not be the right objectives for this audience — a signal that the IN-phase findings may need revisiting.

Navigating Resistance During Rollout

Every program rollout encounters resistance. This is not a sign that the program is weak or that the stakeholder engagement in the IN phase was insufficient. It is a sign that organizations are complex social systems in which change of any kind, including the minor change represented by asking people to attend a learning program, meets the friction of existing priorities, existing relationships, and existing skepticism about whether this intervention will be worth the time it costs.

Experienced designers know this. What they sometimes lack is a systematic approach to diagnosing which type of resistance they are encountering and responding accordingly. The table below maps five resistance types, their signals, and the responses that actually work, which are often different from the responses that feel most natural in the moment.

Resistance Type	The Response That Actually Works
Scheduling resistance *Signal: "We can't fit this in right now."*	This is rarely about the calendar. It is almost always about priority, which is a stakeholder alignment problem that the pilot design needs to solve. Reduce the time commitment required for the pilot cohort, not the program's quality. A thirty-minute pilot experience for eight people generates more useful data than a full program delivered to an audience that attended reluctantly.
Content resistance *Signal: "Our team already knows this."*	Do not argue. Ask what the pilot data shows. If this objection comes before a pilot, respond by building the pilot cohort from the skeptic's team. Either the data confirms their team doesn't need the program (useful to know) or it doesn't (useful to have surfaced). Never defend a program's necessity with logic when you can demonstrate it with evidence.
Ownership resistance *Signal: "Who approved this?"*	This is a governance question wearing a content costume. Someone who should have been engaged in the IN phase was not. The appropriate response is not to defend the program but to bring the person into the process immediately, a review conversation, input on the pilot design, or a role in evaluating pilot results. Resistance that comes from exclusion is resolved by inclusion, not by escalation.
Quality resistance *Signal: "This isn't ready."*	Take this seriously. Distinguish between content that genuinely needs more development and aesthetic preferences that are being elevated to quality concerns. A pilot is explicitly designed to surface this distinction: what doesn't work becomes visible in delivery in a way that design reviews do not reveal. The response to quality resistance is to pilot, not to polish indefinitely.

Cultural resistance *Signal: "This isn't how we do things here."*	The most complex resistance type, and the one most likely to require leadership sponsorship to navigate. Cultural resistance to learning investment is a leadership and change management problem, not a design problem. Escalate to your program sponsor with specific data on what the resistance looks like and what it is costing. Do not try to overcome cultural resistance through better facilitation.

The Perfect vs. Good Enough Conversation

Every designer who takes quality seriously will, at some point in a rollout, face the temptation to delay delivery in order to fix something that is not quite right. The scenario from Event 1 would be stronger with one more revision. The practice activity at Level 2 is slightly too easy. The facilitator guide could be more detailed at the transition between activities 3 and 4. These things are true. They are almost always true. And they are often used, consciously or not, to avoid the exposure of delivery, the moment when the program meets learners and can no longer be protected by the designer's own judgment about its quality.

INDIE is explicit about this: a program that has not been delivered has not been fully designed. The information that delivery generates, what actually lands with learners, what creates confusion, what produces the transfer that the objectives were aiming for, is information that cannot be obtained any other way. Delaying delivery to achieve a higher level of pre-delivery quality is a rational-feeling choice that systematically delays the information that would make the program better.

The practical test for whether something is ready to pilot is not "is it as good as it could be?" It is "is it good enough that the problems it still has won't prevent us from learning what the pilot is designed to teach us?" A scenario that is ninety percent of where it needs to be will still generate useful data on whether the opening works. A practice activity at eighty percent of its target quality will still reveal whether the cognitive level is right. An assessment that could be more polished will still show whether learners can demonstrate the objective.

The version of professional perfectionism that prevents delivery is not a quality commitment. It is a risk-avoidance strategy that consumes the time and organizational capital that iterative improvement requires.

The courage to pilot before the program is finished is part of what the Implementation phase requires.

The Implementation Readiness Test

Before committing to a pilot date, confirm that all of the following are true. If any are not, address them, not because perfection is the standard, but because these represent the minimum conditions for a pilot that will generate useful rather than misleading data.

- The learning objectives are validated and the program directly addresses them. A pilot of a program that addresses the wrong objectives generates data about the wrong design.
- The program has been reviewed by at least one person who was not involved in developing it. Blind spots invisible to the designer are often visible to a fresh reader.
- The facilitator has the materials and has had at least forty-eight hours with them. A facilitator encountering a guide for the first time in the delivery room is not a pilot condition; it is a stress test of the facilitator, not the program.
- A feedback collection mechanism is in place and has been tested. A pilot without feedback collection is a delivery, not a pilot.
- At least one person who is not facilitating will be present to observe. The facilitator cannot simultaneously run the session and watch the learners with the attention quality observation requires.
- The design team has committed to a specific date by which pilot feedback will be synthesized and revisions completed. Without a deadline, the synthesis will expand to fill the available time. ((Continued...))
- Stakeholders have been told what to expect from the pilot: what it will test, how feedback will be collected, and what the improvement process looks like. A stakeholder who expects a polished showcase and encounters a deliberate test is not a stakeholder who has been well managed.

The Example: Ruth's Program Enters Implementation

Ruth's onboarding program had two components: a thirty-minute async eLearning module and a sixty-minute facilitated session. Development had produced materials she was genuinely satisfied with. Implementation began with a pilot of eight new hires, run on a Thursday morning, with Ruth observing and a facilitator who had been with the organization for six years and was well-regarded by the new hire population.

Ruth had briefed the facilitator the previous week. She had also sent a pre-session message to the eight participants four days before, framing the session as: "a conversation about how decisions actually get made here, with people who've recently been where you are." She had written the message with AI assistance and revised it substantially: the AI version had opened with "As part of your onboarding journey, you will attend..." which she removed entirely. The word "journey" had appeared twice.

Three things the pilot revealed that development had not:

1. **The decision matrix was harder to read than it looked on paper.** In development, the matrix had been reviewed by three subject matter experts who all navigated it fluently. In the pilot room, two of the eight participants became visibly uncertain when asked to use it during the Level 1 practice activity. They understood the logic but couldn't find the category they needed quickly enough to maintain confidence. The matrix needed a navigational aid, a column of common decision types cross-referenced to the relevant category. Development had not produced this because the SMEs had not flagged the navigation problem. The SMEs had been using the matrix for years.
2. **The Level 3 scenario produced a productive but unplanned conversation about a gap in the matrix itself.** Two participants identified a category of decision, vendor contract renewals below a certain threshold, that the matrix did not explicitly address. The facilitator handled it well, but Ruth noted that the

omission would surface in every session with operationally experienced new hires. She flagged it for the VP and received a clarification within forty-eight hours. The matrix was updated before the full rollout. This was a content gap that no amount of design review would have found.

3. **The eLearning module was too long for the pre-session assignment.** Participants had been asked to complete the thirty-minute module before the facilitated session. Six of the eight did. Two did not, for reasons that emerged in conversation: both had been in back-to-back meetings the morning before and had not had thirty uninterrupted minutes. Ruth restructured the eLearning into two fifteen-minute segments, with the first becoming the pre-session assignment and the second available as a post-session reference. This changed the facilitated session's Event 3 structure, which had assumed all participants completed the full module, to accommodate participants at different points in the async content. The change took two hours to implement. Finding it had required a pilot.

The one-week feedback sprint ran from the Thursday of the pilot to the following Wednesday. AI synthesized the participant surveys and observer notes on Thursday evening. The design team met for fifty minutes on Friday to categorize and prioritize. Revisions were completed by Monday. The VP received a brief on Tuesday. Full rollout began the following Monday with twenty-two participants across three cohorts, incorporating all three changes.

The full rollout produced Level 1 scores consistently higher than the pilot, and facilitator feedback noted that the navigation aid had eliminated the moment of visible participant confusion that had appeared in the pilot. The matrix update was mentioned positively by three participants who noted that the vendor renewal question had been one they'd been unsure about.

The program that reached the full rollout audience was not the program that came out of Development. It was better. The implementation process had made it better, not by executing the design plan more faithfully, but by treating the first delivery as a source of information rather than a performance of completeness.

What the Implementation Phase Cannot Delegate

The Implementation phase's AI applications are real and useful. But the work that determines whether a rollout succeeds, whether learners arrive in a state of readiness, whether facilitators deliver with confidence and skill, whether managers reinforce rather than undermine transfer, whether resistance is diagnosed and addressed rather than managed around, that work is human, relational, and irreplaceable.

AI can draft the pre-session message. Only the designer can decide whether the tone is right for this relationship with this audience at this moment in the organization's history. AI can produce a facilitator preparation guide. Only someone who knows the facilitator can judge whether the guide addresses the specific things this person is likely to struggle with. AI can synthesize the pilot feedback. Only someone in the room can know that the moment of visible participant confusion during the practice activity was about the matrix's navigation, not the matrix's logic.

Implementation is the phase in which the designer's relational skills are most directly tested: the relationship with the facilitator that makes preparation a collaboration rather than a handoff, the relationship with the stakeholder that allows an honest conversation about what the pilot revealed, the relationship with the learner population that shapes how communications are framed and what organizational signals the program sends about whether learning is taken seriously here.

Sonia's program failed its first launch partly because of logistics, and partly because the implementation had been treated as execution rather than design. The cancelled sessions were a symptom of stakeholder alignment that had not been maintained from the IN phase into deployment. The underprepared facilitator was a symptom of preparation that had been treated as the facilitator's responsibility rather than the designer's. Neither was primarily a logistics failure. Both were relationship failures.

The relational infrastructure of a successful implementation is not something that can be built during the implementation phase. It is built in the IN phase, maintained through Development, and activated during Implementation. A designer who has maintained genuine stakeholder engagement from the first conversation through the final development review is a designer whose stakeholders will answer the phone when the regional manager cancels a session the morning it was supposed to run. That relationship is the most important implementation tool available, and it cannot be prompted into existence.

Chapter Summary

- The launch fallacy treats a program's first delivery as a definitive outcome. INDIE treats it as a source of information: the pilot is a designed learning event for the design team, not a showcase of the program's completeness.
- A well-designed pilot is representative, not optimal: conditions close enough to the full rollout to reveal the problems the rollout will face, small enough to allow rapid adjustment, with a specific observation plan and at least one observer who is not also facilitating.
- The one-week feedback sprint compresses what would otherwise be a multi-week improvement cycle into seven days: capture on Day 1, analyze and categorize on Day 2, revise on Days 3-4, validate on Day 5, communicate on Days 6-7.
- AI applications in Implementation are primarily organizational and communicative: deployment planning, participant and manager communications, facilitator preparation guides, and rapid feedback synthesis. In each case, human judgment about tone, relationship, and organizational context shapes the output.
- Resistance during rollout has five types: scheduling, content, ownership, quality, and cultural. Each type has a specific diagnosis and a response that actually works, which is often different from the response that feels most natural.
- The perfect vs. good enough tension is real, but the version of perfectionism that prevents delivery is risk avoidance, not quality commitment. The test: is the program good enough that its remaining problems won't prevent the pilot from generating useful data?
- What Implementation cannot delegate to AI: the relational infrastructure that determines whether stakeholders stay engaged, whether facilitators are genuinely prepared, whether managers reinforce transfer, and whether resistance is diagnosed accurately. These are human, relational, and irreplaceable.

Reflection Questions

For individual journaling or team discussion.

1. Think of a program rollout that did not go as planned, your own or one you observed. Using the resistance type table in this chapter, which type or types of resistance was the rollout encountering? Was the response to that resistance the right one for its type, or was it a general response applied without diagnosis?
2. Sonia's program met a bad implementation after good design. Have you experienced something similar, work you were proud of that landed badly for implementation reasons? What specifically caused the gap between the quality of the design and the quality of the delivery, and what would a more systematic implementation approach have changed?
3. The chapter argues that delaying delivery to improve quality before piloting is often risk avoidance rather than quality commitment. Do you agree? Have you been in situations where you or a colleague delayed a pilot for quality reasons that, in retrospect, were actually avoidance? How do you tell the difference in the moment?
4. Apply the Implementation Readiness Test to a program currently in development or recently launched. Which conditions were fully met? Which were not? What was the consequence of the conditions that were not met, and what would it have taken to meet them?

Part Two · Chapter 9

Phase E

Enhancement: From Evaluation to Continuous Improvement

The evaluation report was excellent. Forty-seven pages, professionally formatted, organized across Kirkpatrick's four levels with data at each. Level 1 satisfaction scores averaged 4.3 out of 5. Level 2 assessment results showed an average improvement of 31 percentage points from pre- to post-test. Level 3 data, collected ninety days after the program, indicated that 74 percent of participants reported applying the learning in their work. Level 4 data linked the program to a measurable improvement in the quality metric it had been designed to address.

Thomas had written it himself. He was proud of it. He presented it to the leadership team in October, eighteen months after the program had launched. The leadership team received it well. There were questions about methodology. There was appreciation for the rigor. There was a round of applause, which Thomas had not expected and which he found faintly uncomfortable.

Then someone asked: "Can we scale this?"

Thomas had not prepared for that question. He had spent eighteen months preparing to prove that the program worked. He had not spent eighteen months preparing to make it better. The evaluation report told the leadership team what had happened. It told them very little about what to do next.

The Level 3 data, that 74% of participants reported applying the learning, was the most consequential number in the report. Twenty-six percent had not. Thomas did not know why. He did not know which participants, which competencies, which job contexts had produced the non-transfer. He did not know whether the gap was a design problem, a manager problem, or a systems problem. The data had been collected once, ninety days after the program, and it had been collected to produce the report, not to improve the program.

The applause faded. The leadership team moved on to the next agenda item. Thomas went back to his desk and began, for the first time, to think about what evaluation was actually for.

What Evaluation Has Always Been For

Donald Kirkpatrick began publishing his four-level evaluation model in 1959. For sixty-five years it has provided the field of training and development with its primary framework for thinking about whether learning programs work. It is genuinely valuable, and the fact that it has survived six decades of academic critique, competing models, and technological transformation is evidence of something real. The four questions it poses: Did learners react positively? Did learning occur? Did behavior change? Did results improve? These are the right questions.

The problem is not Kirkpatrick's model. The problem is what happened to it in practice. The four levels became, in most organizations, a justification structure rather than an improvement mechanism. Level 1 data was collected because it was easy and stakeholders expected it. Level 2 data was collected because it produced numbers. Levels 3 and 4 were attempted periodically, when resources allowed, as evidence that the L&D function was contributing to organizational outcomes. The data arrived after the program was over. It was used to write reports. Those reports were used to justify the next budget cycle. And then the cycle began again, with the same programs, largely unchanged.

This is not what Kirkpatrick designed his model to do. He designed it to answer the question: how do we know whether training is working, and how do we use that knowledge to make it work better? The second half of that question, the improvement half, got lost somewhere between the model's creation and its widespread adoption. What remained was the measurement without the improvement. Evidence without action.

INDIE's Enhancement phase recovers the second half of the question. It reframes evaluation not as a reporting activity that happens after programs are delivered, but as a continuous information process that runs alongside delivery, feeds back into design, and produces programs that improve iteratively rather than aging statically. The framework is Kirkpatrick's. The relationship to time, and to action, is fundamentally different.

> *Evaluation without action is just measurement. Enhancement is what happens when measurement drives the next design decision.*

The Shift From Evaluation to Enhancement

The word "evaluation" carries a terminal connotation. You evaluate something that is finished, a decision that has been made, a project that has been completed, a program that has been delivered. Evaluation looks backward at what happened and reaches a verdict about whether it was good.

The word "enhancement" carries a generative connotation. You enhance something that is continuing, something that has value now and could have more value with the right adjustment. Enhancement looks at what is happening and asks what could be better. It does not wait for the program to be over. It operates continuously, treating every delivery as both an outcome to be measured and a source of information to be used.

This is not a semantic distinction. It changes when data is collected, what data is collected, who receives it, and what happens as a result. In a traditional evaluation model, data collection is episodic: Level 1 surveys after each session, Level 2 assessment results compiled periodically, Level 3 follow-up surveys conducted once at a fixed interval after the program. In INDIE's Enhancement phase, data collection is continuous: LMS analytics running in real time, participant survey results synthesized within twenty-four hours of each session, performance data monitored on a defined cadence from the first day of rollout, manager feedback incorporated as a routine part of the program's ongoing maintenance.

The practical enabler of this shift is AI. The continuous data streams that modern learning environments generate have always contained more improvement signal than teams could extract manually. AI provides the analytical capacity to process those streams in near-real-time, surfacing patterns that would take weeks to find manually and flagging issues that would otherwise persist through many more delivery cycles before anyone noticed them.

Theoretical Grounding: Systems Thinking and the Learning Organization

Peter Senge's work on the learning organization, and the broader systems thinking tradition from which it draws, provides the conceptual foundation for the Enhancement phase that Kirkpatrick's measurement model does not supply on its own. A learning organization is one that systematically learns from its own experience, adjusting its practices in response to feedback loops that connect outcomes to causes to adjustments to new outcomes. INDIE's Enhancement phase is the application of this principle to the learning programs themselves: programs become learning organizations in miniature, continuously processing feedback about their own effectiveness and improving in response. The feedback loops are the mechanism. AI is what makes them fast enough to be useful. Human judgment is what determines whether the adjustments they suggest are the right ones.

Kirkpatrick's Four Levels in the Enhancement Phase

Kirkpatrick's four levels do not disappear in INDIE. They are restructured. Each level becomes a continuous data stream rather than a periodic measurement event, and each stream is connected to a specific improvement action rather than a reporting output. What follows is each level treated with the fresh eyes that the Enhancement frame requires: not "what does this level measure?" but "what does this level tell us that we can act on, and how quickly can we get that information?"

Level 1 **Reaction:** *Did they find value in it?*

Level 1 is the most collected and least acted-on evaluation data in the field. The typical pattern: satisfaction surveys are administered after every session, scores are compiled monthly, averages are reported quarterly, and the information influences future program design in ways that are rarely traceable. In the Enhancement phase, Level 1 data is collected after every session and synthesized within twenty-four hours, not to monitor satisfaction as a proxy for quality, but to identify specific, actionable signals about what is and isn't working for this audience in this delivery context. The question is not "what was the average score?" It is "what specific element generated the lowest scores, and what does the open-ended feedback tell us about why?" The answer to that question drives a specific design adjustment, not a general report.

AI Applications

AI synthesizes open-ended Level 1 responses within hours of each session, identifying themes, flagging language that signals confusion or disconnection, and producing a prioritized list of actionable signals. Sentiment analysis on open-ended responses surfaces the emotional register of the feedback, not just what participants said but how they said it, which often carries information that quantitative scores do not.

Level 2 Learning: *Did competence develop?*

Level 2 assessment data is the most direct measure of whether the program accomplished its instructional purpose. In the Enhancement phase, this data is examined at two levels: aggregate performance against each objective (which objectives are most learners meeting? which are they consistently missing?) and item-level analysis (which specific assessment items are producing high error rates, and what do those error patterns suggest about where the instruction is falling short?). An objective that consistently produces low assessment scores is a signal that the Design or Development phase needs revisiting for that specific competency. An assessment item that consistently attracts a specific wrong answer is a signal that a specific misconception is not being addressed by the instruction.

AI Applications

AI analyzes assessment response patterns across cohorts, identifying items with anomalous error rates, surfacing which distractors are most frequently chosen and what that reveals about learner misconceptions, and producing objective-level performance summaries that connect assessment data to specific instructional elements. This analysis, done manually across multiple cohorts, requires significant time. AI produces it continuously.

Level 3 Behavior: *Did performance change on the job?*

Level 3 is where most evaluation efforts either stop or produce data too late and too general to act on. The ninety-day survey, collected once, compiled into a percentage ("74 percent report applying the learning"), and filed in the evaluation report is the archetypal Level 3 dead end. In the Enhancement phase, Level 3 data collection is designed from the start to be specific and actionable: which competencies are transferring? which are not? which job contexts are supporting transfer? which are not? which participants are applying the learning? which are not, and why? The answers to these questions direct specific improvements: to Event 9 design if the transfer infrastructure is insufficient, to the manager briefing if managerial support is absent, to the objectives themselves if what was taught does not match what the job requires.

AI Applications

AI analyzes Level 3 survey responses and performance observation data to identify transfer patterns by competency, cohort, manager, and job context. Natural language processing on open-ended transfer survey responses surfaces the specific barriers participants are encountering, language the quantitative transfer scores do not capture. Predictive modeling can identify the learner and context characteristics most associated with successful transfer, informing both future selection of pilot cohorts and manager briefing design.

Level 4 Results: *Did it move the needle on what matters?*

Level 4 is the most important level and the least often reached, because connecting training outcomes to business results requires patience, organizational access, and methodological care that most L&D teams are not resourced to provide. INDIE does not pretend this is easy. It does insist that it is necessary, not for every program, but for programs that are significant enough to warrant the investment. The prerequisite for useful Level 4 data is Level 4 thinking in the IN phase: if the business metric the program is meant to move was not identified before design began, it cannot be measured after the program runs. Enhancement makes Level 4 measurement viable by connecting it to the stakeholder alignment established in the IN phase: the metrics were agreed before the program was designed, the measurement methodology was defined before the program was built, and the timeline for expected behavior change was negotiated before the program was launched. This preparation is what makes Level 4 data available rather than aspirational.

AI Applications

AI analyzes business performance data at the defined cadence, identifies correlations between program completion and metric change, and controls for confounding variables to the extent the available data allows. Executive reporting for Level 4 findings, the most demanding communication task in the Enhancement phase, can be drafted by AI and refined by the designer for the specific audience, relationship, and organizational context.

Building the Enhancement Infrastructure

The Enhancement phase does not run automatically. It requires infrastructure: the data systems to collect continuous feedback, the analytical processes to make that feedback actionable, the organizational rhythms to review and act on the findings, and the design authority to implement improvements without waiting for a formal program revision cycle. Building this infrastructure is work that must happen alongside the IN and Development phases, not after the program has launched.

The Enhancement Dashboard

Every program in INDIE should have an Enhancement dashboard, a single view of the data streams that matter most for this program's improvement, designed for the people who will use it to make decisions. The dashboard is not a reporting tool. It is a decision-support tool, and the distinction matters: a reporting tool shows what happened. A decision-support tool shows what to do next.

Effective Enhancement dashboards have three design properties. They are audience-appropriate: a dashboard designed for the instructional designer contains different data at different levels of granularity than one designed for the program sponsor, which contains different data than one designed for the facilitator reviewing their own session. They are action-oriented: every metric on the dashboard should be connected to a specific action that a specific person can take if the metric falls below a defined threshold. And they are updated continuously, not periodically: a dashboard that shows last quarter's data is a history lesson. A dashboard that shows last week's data is a management tool.

PROMPT: Enhancement Dashboard Design
I need to design an Enhancement dashboard for [PROGRAM TITLE]. Program description: [BRIEF SUMMARY OF PROGRAM AND OBJECTIVES] Primary stakeholders who will view the dashboard: [LIST BY ROLE] Data sources available: [LMS DATA / SURVEY TOOL / HRIS / PERFORMANCE SYSTEM] Update frequency achievable: [REAL-TIME / DAILY / WEEKLY] Business metric this program is designed to influence: [DESCRIBE] Please design a dashboard specification that includes: 1. A designer view with: - Completion rate by cohort and session - Level 2 assessment performance by objective - Level 1 satisfaction scores with open-ended themes - Level 3 transfer rate by competency (at defined follow-up interval) - Flagged alerts: items falling below defined thresholds

2. An executive/sponsor view with:
   - Program reach (participants completed vs. total target)
   - Headline competency achievement rate
   - Level 4 business metric trend (if data available)
   - One key insight and one recommended action from current data

3. A facilitator view with:
   - Their session-level Level 1 scores vs. program average
   - Specific open-ended feedback from their sessions
   - Suggested adjustments based on their session data

4. Alert thresholds: define the specific metric values that should trigger a review conversation, with the recommended action for each alert

Format as a dashboard specification document, not a prototype.

The Enhancement Cadence

Infrastructure without rhythm is a filing cabinet. The Enhancement phase requires a defined cadence of data review, decision-making, and action, a recurring organizational rhythm that treats improvement as a standing commitment rather than an occasional project.

The minimum viable Enhancement cadence for a program in active rollout has three components. A weekly session review: within twenty-four to forty-eight hours of each delivery, Level 1 data is synthesized and reviewed, and any delivery-level adjustments that can be made before the next session are made. A monthly program review: Level 1 and Level 2 trends are examined across all cohorts to date, patterns are identified, and design-level adjustments are scoped and prioritized. A quarterly impact review: Level 3 and Level 4 data is examined, the program's contribution to the business metric is assessed, and the program sponsor receives a briefing that connects the program's performance to the organizational goals it was designed to serve.

These are meetings that need to happen, with specific people, on a specific schedule, with specific decision authority. The most common reason Enhancement cycles stall is not that the data is unavailable. It is that no one has been assigned the responsibility to review it on a schedule, and no one has been given the authority to act on what it reveals without seeking approval.

The Enhancement Report

Different audiences need different views of the same Enhancement data, and the wrong view presented to the wrong audience produces at best confusion and at worst the conclusion that the data doesn't support the program. A designer reviewing item-level assessment data needs granularity that would overwhelm a CFO. A CFO reviewing Level 4 business results needs a narrative connection to organizational goals that a designer's technical analysis does not provide. AI can produce both from the same underlying data, and the designer's role is to verify that each version accurately represents what the data actually shows.

PROMPT: Kirkpatrick Four-Level Analysis
I need to analyze Enhancement data for [PROGRAM TITLE] across all four Kirkpatrick levels. Level 1 data: [PASTE QUANTITATIVE SCORES AND OPEN-ENDED RESPONSES] Level 2 data: [PASTE ASSESSMENT PERFORMANCE BY OBJECTIVE AND ITEM] Level 3 data: [PASTE TRANSFER SURVEY RESPONSES AND/OR OBSERVATION DATA] Level 4 data: [PASTE BUSINESS METRIC DATA AND BASELINE] Number of participants across all cohorts: [NUMBER] Program objectives: [PASTE] Business metric this program was designed to influence: [DESCRIBE] Please produce: 1. An analysis at each Kirkpatrick level: - What the data shows (key patterns, not raw numbers) - What it suggests about program effectiveness - The specific design or delivery element most likely to explain the pattern - The recommended action 2. Cross-level correlations: - Where high Level 1 scores correlate or fail to correlate with

Level 2 outcomes
- Where Level 2 competency achievement correlates or fails to correlate with Level 3 transfer
- Any pattern that suggests the evaluation data may be measuring the wrong things

3. A priority-ranked list of 3 improvements, with the rationale for each ranking based on impact potential and implementation effort

4. A one-paragraph executive summary suitable for a program sponsor who will have 90 seconds to read it before a leadership meeting

PROMPT: Executive Enhancement Report

I need to write an executive report on [PROGRAM TITLE] for [AUDIENCE: e.g., VP of Operations, HR Leadership Team, Board L&D Committee].

Program purpose: [ONE SENTENCE]
Total participants to date: [NUMBER]
Time period covered: [DATE RANGE]
Key findings from Kirkpatrick analysis: [SUMMARIZE FROM PRIOR ANALYSIS]
Business metric movement: [DESCRIBE: INCLUDE BASELINE AND CURRENT]
Investment made: [APPROXIMATE DESIGN AND DELIVERY COST IF AVAILABLE]
Key improvements made since launch: [LIST 2-3]

Please produce an executive report that:
- Opens with one sentence: the program's business purpose and current status
- Presents 3-4 key findings as headlines with one supporting sentence each
- Connects competency outcomes to business metric outcomes explicitly
- Acknowledges one area where results are not yet where they should be and what is being done about it
- Closes with a specific recommendation and requested decision
- Fits on two pages maximum

Tone: Confident and data-grounded, not defensive. Honest about gaps, clear about progress. Assumes an audience that respects evidence and has limited patience for L&D jargon.

When to Tune, When to Replace

One of the most consequential judgments in the Enhancement phase is the distinction between a program that needs tuning and a program that needs replacing. These are not the same decision, and treating one as the other wastes resources in one direction or misapplies them in the other.

A program that needs tuning has the right design responding to the right problem for the right audience, and specific elements of its implementation or design that can be improved without rethinking the program's fundamental architecture. Tuning is the default mode of the Enhancement phase: iterative improvements to specific elements based on specific data.

A program that needs replacing has a more fundamental problem: the objectives are no longer the right objectives, the audience has changed in ways the design cannot accommodate, the organizational context has shifted in ways that make the program's assumptions obsolete, or the root cause analysis that justified the program in the first place has been superseded by new information. Tuning a program that should be replaced is one of the most common and most costly mistakes in L&D practice. It produces a program that is incrementally better at being the wrong program.

The table below maps Enhancement data signals to the appropriate response. It is a decision tool, not a formula, the signals indicate, but human judgment about organizational context, resource availability, and strategic priority determines the final decision.

Signal from Enhancement Data	Tune the Program ✓	Replace the Program ↺
Completion rates are low	Reduce length, improve access, check scheduling friction	If completion has been persistently low across multiple iterations despite tuning, the program may not have organizational priority, a leadership problem, not a design one
Level 1 satisfaction is low	Review facilitation, pacing, scenario relevance, and audience-content match	If satisfaction has improved through iteration but remains low, the audience and

		the program may be fundamentally misaligned in ways that redesign cannot resolve
Level 2 assessment scores are low	Review cognitive load, practice design, feedback quality, and objective-assessment alignment	If scores remain low after multiple rounds of Development improvement, the objectives themselves may be unrealistic for the audience and available learning time
Level 3 transfer is not occurring	Add manager briefings, strengthen Event 9, improve transfer design, address specific environmental barriers	If transfer has never occurred despite strong Level 1-2 data and explicit transfer design, the performance gap may not be primarily training-addressable, return to the IN-Phase root cause analysis
Level 4 business results are unchanged	Review whether the right metrics are being measured and whether enough time has elapsed for behavioral change to produce measurable results	If Level 3 transfer is confirmed but Level 4 results remain unchanged after sufficient time, the training objective may be addressing the wrong driver of the business outcome
Content is significantly out of date	Update specific content sections, verify objectives still reflect current performance requirements, revise examples and scenarios	If the foundational context has changed, new regulatory environment, organizational restructuring, technology platform change, tuning is patching. Start the IN Phase again.

The Enhancement Cycle: Priya's Program Over Time

The INDIE framework is not a process you complete once and then start over. It is a cycle that a program enters and moves through continuously. What follows traces Priya's onboarding program through three Enhancement cycles, eighteen months of delivery, measurement, and improvement, to show what continuous enhancement looks like in practice, at a concrete level.

Month 1–2	**Launch + Immediate Enhancement** Full rollout delivers to 68 new hires across seven cohorts. Level 1 scores average 4.1/5. Level 2 shows strong performance on Objectives 1 and 2 (organizational context); Objective 3 (decision matrix navigation) averages 71%, below the 80% threshold. AI surfaces a consistent

	pattern in the Level 1 open-ended responses: participants feel confident about the matrix but uncertain about edge cases in their specific department. Priya adds two department-specific worked examples to the eLearning module and revises the Level 3 facilitated scenario to include a cross-departmental edge case. Second-cohort Level 2 scores for Objective 3 improve to 79%.

Month 3	**First Transfer Follow-Up** Level 3 survey administered to Month 1-2 cohort (n=68, response rate 74%). Transfer rate on Objective 3 is 61%, lower than Objectives 1 and 2 at 82% and 78% respectively. AI analyzes the open-ended responses and surfaces a specific barrier: new hires are confident using the matrix but hesitant to initiate the consultation process because they worry about appearing incompetent to their manager. This is a manager culture signal, not a design signal. Priya revises the manager transfer guide to explicitly name this dynamic and adds a coaching question: "Has your new hire come to you with a question about a decision boundary this month? If not, that's worth asking about."

Month 4–6	**Design Revision Cycle** Monthly program review surfaces a pattern in Level 1 data across cohorts: the sixty-minute facilitated session is running long in four of the seven department cohorts due to the cross-departmental scenario generating extended discussion. This is a facilitation constraint, not a design problem, but it is causing timing pressure that is compressing Event 9 retention activities in some sessions. Priya separates the cross-departmental scenario into an optional extension activity, making it available for cohorts with strong discussion engagement and preserving Event 9 time for all cohorts. She also adds a ten-minute "facilitator flex" note to the guide with specific guidance on when to use it.

Month 9	**Quarterly Impact Review** Level 4 data arrives from the performance analytics team. The quality metric the program was designed to address has improved 14% across the departments with the highest program completion rates, compared to 3% in departments with lower completion. The correlation is not causation, and Priya says so in the report. But the pattern is consistent with the program's theory of change and provides grounds for a meaningful conversation with the VP about expanding rollout to the remaining departments that have not yet participated. She presents the data with appropriate methodological caveats and a specific recommendation: expand rollout, add a manager briefing requirement for departments with completion rates below 70%.

Month 12–18	**Program Maturity and Content Update** The organizational decision matrix itself has been updated. A restructuring has changed the approval thresholds for two decision categories. Priya identifies this as a content update requirement, not a redesign trigger. She updates three eLearning screens, revises two assessment items, and sends a targeted refresher to all participants from the prior twelve months who are now operating under the updated thresholds. The Enhancement dashboard flags which prior participants have completed the refresher and which have not, allowing coordinator follow-up on the low-completion departments. By month 18, the program has been delivered to 247 new hires across all departments. Its design has been improved in six specific, data-driven ways since launch. It is demonstrably better than it was on the day it first ran. That is what Enhancement looks like.

Thomas, Revisited

Thomas eventually answered the question about scaling. Not immediately: it took him several months, and a new project, and a different relationship with what evaluation was for.

The new project was a leadership development program, smaller in scope than the one he had spent eighteen months evaluating. He started collecting data differently. Not to build a report. To build a program.

He analyzed Level 1 responses after every session, looking not for the average score but for the specific signal in the open-ended feedback. He examined Level 2 item performance after every cohort, looking not for the aggregate pass rate but for the specific misconceptions the wrong answers revealed. He designed Level 3 follow-up at thirty and sixty days, and he designed the questions specifically to tell him whether the transfer gap was a design problem or a manager problem. He built a dashboard that his program sponsor could read in three minutes and his facilitator could read in ninety seconds.

When the leadership team asked about this program eighteen months in, he was ready for the question about scaling. He knew what was working because he had been measuring it continuously. He knew what had been fixed because he had fixed six things. He knew what was still not right because the data was still telling him. He did not produce a forty-seven-page report. He produced a two-page brief with four findings, three improvements made, and a specific recommendation. There was no applause. There was a decision. The program was scaled.

That is the difference between evaluation and Enhancement. Evaluation produces applause. Enhancement produces decisions. And decisions are what learning programs exist to support.

Chapter Summary

- Kirkpatrick's four levels are the right questions. What went wrong in practice was that they were disconnected from the improvement actions they were always meant to drive, and that data collection happened after programs were over rather than continuously alongside delivery.
- Enhancement replaces evaluation by restructuring the relationship to time and to action: data collection is continuous, analysis is near-real-time, and findings connect directly to specific improvement actions rather than to periodic reports.
- AI provides the analytical capacity that makes continuous Enhancement operationally feasible: synthesizing open-ended feedback within hours, identifying assessment item patterns across cohorts, surfacing transfer barriers from narrative survey data, and drafting audience-appropriate reports from the same underlying data.
- The Enhancement infrastructure has three components: a dashboard designed for decision support rather than reporting, a cadence of weekly, monthly, and quarterly reviews with defined decision authority, and audience-appropriate reporting that connects different levels of data to different levels of organizational decision-making.
- The tuning vs. replacing judgment is the most consequential decision in the Enhancement phase. Tuning a program that should be replaced is one of the most common and most costly mistakes in L&D practice. The signal table in this chapter maps data patterns to responses, but human judgment about organizational context determines the final call.
- The Enhancement cycle is not a phase that ends. It runs continuously alongside and beyond delivery, producing programs that are demonstrably better after eighteen months than they were on the day they launched.
- What AI cannot do in the Enhancement phase: determine whether a correlation between program completion and metric change implies causation, judge whether a specific improvement is worth the organizational capital it will cost to make, or tell a program sponsor something they need to hear in a way that preserves the relationship. Those are irreducibly human judgments.

Reflection Questions

For individual journaling or team discussion.

1. Think of a program evaluation report you have produced or received. What specific design improvements did it produce? If the answer is "none" or "unclear," what would it have taken for the same data to drive a specific improvement decision?
1. Of Kirkpatrick's four levels, which one generates the most data in your current organization and which generates the least? What organizational factors explain that pattern, and which of those factors are within your influence to change?
2. Apply the tuning vs. replacing table to a program currently in your portfolio. Based on the Enhancement data you have or could have, what does the evidence suggest? If the evidence points toward replacing, what would it take to make that recommendation to the relevant stakeholders?
3. Thomas's shift was from building data to justify a program to building data to improve a program. Is there a program in your current work where you are primarily in justification mode rather than improvement mode? What would it take to shift?
4. The chapter argues that Enhancement requires a defined cadence with specific decision authority, that data without organizational rhythm is a filing cabinet. Looking at your current programs, who has the authority to make design changes based on Enhancement data without seeking additional approval? If the answer is unclear, what is the consequence of that ambiguity?

Part Three · Chapter 10

The Instructional Designer and the Learning Experience Designer

They sat on opposite sides of the same table, which was appropriate, because they were coming at the same problem from opposite directions.

Evan had been an instructional designer for eight years. He thought in objectives and outcomes, in cognitive levels and assessment alignment, in the logical architecture of how a learning experience should be structured so that a person moved from not knowing to knowing and from knowing to doing. He was rigorous and systematic and occasionally drove his colleagues slightly mad by insisting on Mager format for every objective before a single piece of content was written.

Francis had been a learning experience designer for five years, arriving in L&D by way of UX design and a degree in human-computer interaction. She thought in journeys and emotions, in friction points and moments of delight, in the feeling of a learner encountering a new idea and whether that feeling was curiosity or confusion or the particular low-grade dread that compliance training reliably produced. She was perceptive and empathic and occasionally drove her colleagues slightly mad by asking "but how does it feel?" about things that seemed like they should be purely structural.

The project in front of them was a new product knowledge program for a customer-facing sales team. It had a hard deadline, a demanding stakeholder, and, in Evan and Francis's shared assessment, a previous iteration that had failed not because it was inaccurate but because nobody had wanted to use it. The content had been correct. The experience had been miserable. People had clicked through it as fast as possible, retained almost nothing, and gone back to doing what they had always done.

They spent the first two hours of the project doing what they always did when they worked together: arguing productively about which problem to solve first. Evan wanted to establish the objectives before anything else. Francis wanted to understand why the previous program had produced dread before they designed anything. Both were right. That was, as it turned out, the whole point.

Two Roles, One Program

The distinction between instructional design and learning experience design is one of the most debated and least resolved questions in the contemporary L&D field. Some organizations treat them as synonymous, using the titles interchangeably for the same role. Others treat them as entirely separate disciplines with different tools, different vocabularies, and different organizational homes. Neither position fully captures the relationship.

They are distinct disciplines with meaningfully different orientations, and they are most powerful when they work together. The instructional designer's orientation is toward learning architecture: the systematic structure that moves a learner from a current state to a target state, grounded in learning theory and driven by clearly specified objectives. The learning experience designer's orientation is toward the learner's subjective encounter with that architecture: how it feels to move through it, where it creates delight or confusion or the particular frustration of not being able to find what you need when you need it.

A program designed only by the ID orientation is often technically correct and experientially deadening, the content is accurate, the sequence is logical, the assessments are valid, and nobody wants to be in it. A program designed only by the LXD orientation is often beautifully crafted and pedagogically hollow, the interface is lovely, the journey feels coherent, and the learner arrives at the end having been through an experience without quite having learned anything. The programs that achieve both, that are architecturally sound and experientially compelling, are the ones where both orientations are genuinely present, either in one practitioner who holds both or in a collaboration between practitioners who each bring one.

Instructional Designer (ID)	Learning Experience Designer (LXD)
Primary question What must the learner be able to do, and how do we systematically build that capability?	**Primary question** What does the learner experience at each moment, and how does that experience serve or undermine the learning?
Core discipline Learning theory, cognitive science, performance analysis, systematic curriculum design	**Core discipline** UX design, human-computer interaction, empathy research, narrative and emotional arc design
Primary deliverables Objectives, curriculum maps, storyboards, assessment instruments, facilitation guides	**Primary deliverables** User journey maps, persona-based experience audits, interaction prototypes, accessibility reviews, feedback architecture
Risk of working alone Architecturally rigorous programs that are experientially flat, confusing to navigate, or emotionally disconnected from learners	**Risk of working alone** Beautifully crafted experiences with insufficient pedagogical structure, assessment, or transfer design
In the INDIE framework Leads IN-phase objective development and curriculum architecture; drives Development through Gagné framework; defines Enhancement metrics	**In the INDIE framework** Leads IN-phase learner analysis and persona development; leads UX and emotional arc design in Development; drives accessibility throughout
AI augmentation focus Objective generation and critique, content chunking, assessment design, Bloom's alignment, Kirkpatrick analysis	**AI augmentation focus** Persona development, user journey mapping, accessibility auditing, emotional arc design, friction point identification

The Instructional Designer in INDIE

If you are an instructional designer reading this chapter, much of Part Two was written with you in mind. The INDIE framework's theoretical foundations: Knowles, Gagné, Mager, Vygotsky, Bloom, Kirkpatrick, are the frameworks your professional identity is built around. The IN-Phase emphasis on performance analysis, gap identification, and objective precision reflects the core of ID practice. The Development phase's Gagné architecture is the framework you likely internalized early in your career and have been applying, with varying levels of explicitness, ever since.

What INDIE changes for the instructional designer is not the theoretical foundation. It is the operational layer: the speed at which analysis and design can be run concurrently, the AI tools that accelerate the drafting work that once consumed the majority of project time, and the Enhancement infrastructure that transforms evaluation from a reporting activity into an improvement mechanism. The principles are yours. The tools are new.

There is also something INDIE asks of the instructional designer that the traditional ID role did not ask as directly: a genuine reckoning with the learner's subjective experience. Gagné's nine events describe the psychological conditions for learning, but they do not describe what it feels like to move through a program, whether the opening scenario creates genuine curiosity or dutiful compliance, whether the practice activities produce productive challenge or discouraging confusion, whether the assessment feels like an opportunity to demonstrate competence or a threat to professional self-concept. INDIE integrates the LXD orientation not as an optional add-on but as a design requirement. For instructional designers who have not previously worked in partnership with LXD colleagues, this may require developing new muscles.

Where AI Changes the ID Role Most

The tasks that have historically consumed the most instructional designer time, writing and revising objectives, drafting content outlines, developing assessment items, producing facilitation guides, are precisely the tasks that AI handles most effectively. This changes the economics of ID work in a specific way: the ratio of design time to production time shifts dramatically toward design.

An instructional designer who spent sixty percent of their project time on production tasks, drafting, formatting, revising, producing, and forty percent on design tasks, analysis, judgment, stakeholder navigation, quality evaluation, will find, in an AI-augmented workflow, that the ratio inverts.

Production tasks that took two days take two hours. The time that is freed does not disappear into efficiency gains for the organization alone. It becomes available for the design work that AI cannot do: the stakeholder relationship management that keeps a program aligned with organizational reality, the pedagogical judgment that evaluates AI output against genuine learning standards, the transfer design that connects the training environment to the job environment, and the Enhancement analysis that makes programs better over time.

The risk is that this freed time gets absorbed into the same production work at higher volume rather than invested in deeper design work. Organizations that respond to AI efficiency gains in instructional design by expecting the same designer to produce twice as many programs in the same time are not investing in better design. They are investing in more production, and the quality consequences will arrive in the Enhancement data.

Role Callout: Instructional Designer

The most important professional development investment for an ID in an AI-augmented environment is not learning to use AI tools, that's table stakes, and the tools themselves are learnable quickly. The most important thing is developing the evaluative judgment that makes AI output usable: the ability to look at a set of AI-generated objectives and know which ones are at the wrong Bloom's level, which scenarios are missing the organizational specificity that makes them authentic, which assessment items are measuring recall rather than the application they claim to target. That judgment is built through deliberate practice across many projects, not through tool adoption. The designer who invests in developing it becomes more valuable as AI tools proliferate, not less.

The ID's AI Prompt Toolkit

The prompts in Chapters 6 through 9 cover the full INDIE workflow. What follows are three prompts specific to the ID role's distinctive contribution: performance consulting at the front end, the kind of pedagogical quality review that only expertise can perform, and the learning-to-performance gap analysis that connects instruction to outcomes.

PROMPT: Performance Consulting: Is Training the Right Solution?

I have received a request for training on [TOPIC] for [AUDIENCE]. Before designing anything, I need to assess whether training is the right intervention.

Here is what I know about the performance gap:
Stated problem: [DESCRIBE WHAT WAS COMMUNICATED]
Current performance: [DESCRIBE IF KNOWN]
Desired performance: [DESCRIBE IF KNOWN]
Available context: [PASTE ANY RELEVANT DOCS, DATA, OR NOTES]

Please analyze this request against Gilbert's Behavior Engineering Model, classifying potential causes across six categories:

1. Data/information (do people know what is expected and how they are doing?)
2. Resources/tools (do people have what they need to perform?)
3. Incentives/consequences (are the right behaviors reinforced?)
4. Knowledge/skills (do people know HOW to perform?)
5. Capacity (are people physically/cognitively able to perform?)
6. Motives/preferences (do people want to perform?)

Based on this analysis:

- What proportion of the gap is likely training-addressable (Category 4)?
- What non-training interventions should be considered alongside or instead?
- What additional information would significantly change this assessment?
- What should I say to the stakeholder if training is not the primary solution?

PROMPT: Pedagogical Quality Review of AI-Generated Content

I need to conduct a pedagogical quality review of the following AI-generated content for [MODULE TITLE]:

[PASTE AI-GENERATED CONTENT]

Validated learning objective this content serves: [PASTE OBJECTIVE]
Target audience: [DESCRIBE]
Bloom's level required: [SPECIFY]

Please evaluate this content against the following criteria:

1. Objective alignment: Does this content directly serve the stated objective, or does it address adjacent content that is accurate but not required?

2. Cognitive level: Does the content develop the cognitive operation the objective requires, or does it develop a lower-level operation instead?
3. Cognitive load: Does this content exceed working memory capacity? Identify any sections with more than 5 distinct concepts.
4. Schema connection: Does this content connect to prior knowledge that the learner analysis suggests this audience has? Or does it assume foundational knowledge that may not be present?
5. Example quality: Are the examples specific to the [INDUSTRY/ROLE] context, or are they generic enough to feel inauthentic?
6. Extraneous content: What, if anything, could be removed without reducing the content's ability to serve the objective?

Provide a revised version of any section that fails these criteria, with brief annotations explaining each change.

PROMPT: Learning-to-Performance Gap Analysis
I have Level 2 and Level 3 Enhancement data for [PROGRAM TITLE] and need to analyze the gap between what learners achieve in training and what they demonstrate in performance. Level 2 data (assessment outcomes by objective): [PASTE DATA] Level 3 data (transfer survey / observation data): [PASTE DATA] Job context in which transfer should be occurring: [DESCRIBE] Known transfer barriers from prior research: [DESCRIBE IF AVAILABLE] Please analyze: 1. Which objectives show strong Level 2 achievement but weak Level 3 transfer? (Competence achieved in training but not appearing in performance) 2. Which objectives show weak Level 2 AND weak Level 3? (Competence not achieved in training or performance) 3. For Category 1 (learned but not transferred), classify the likely barrier: - Transfer-inappropriate practice (training conditions too different from job) - Environmental barriers (tools, time, support missing on the job) - Motivational barriers (no incentive or reinforcement for new behavior) - Retention failure (learning occurred but decayed before application) 4. For each barrier type, the specific design change most likely to address it Flag any pattern suggesting the objectives themselves may need revision rather than the instructional design that serves them.

The Learning Experience Designer in INDIE

If you are a learning experience designer reading this chapter, you may have found Part Two useful but somewhat lopsided toward the ID orientation, heavy on learning theory and instructional architecture, lighter on the experiential and emotional dimensions that define your distinctive contribution. That imbalance is partly by design: Part Two establishes the framework's theoretical foundations, which are primarily grounded in the ID tradition. This section redresses it.

The learning experience designer's core insight, that the learner's subjective experience of a program is not decorative but pedagogically consequential, is one that the instructional design field has been slow to fully absorb. A learner who feels confused, dismissed, or talked down to is not in a cognitive state that supports learning. A learner who feels genuinely seen by the scenario in Event 1, appropriately challenged by the practice in Event 6, and respected by the feedback in Event 7 is in a state that does. Emotional design is not ornamentation applied to instruction. It is part of what makes instruction work.

The LXD's INDIE work is distributed across all four phases but is most concentrated in three specific areas: the learner analysis stream of the IN phase, the emotional arc and UX design dimensions of the Development phase, and the accessibility architecture that runs through the entire program lifecycle. Each of these is addressed in turn.

LXD in the IN Phase: Learner Analysis and Empathy Research

The persona development work described in Chapter 6 is primarily LXD territory. An ID-only persona describes who the learner is and what they need to know. An LXD-informed persona also describes how the learner feels about the learning itself, whether they arrive with anticipation or dread, whether the topic touches on professional identity in ways that create defensiveness, whether the delivery format creates friction that the content would not otherwise produce.

These emotional and experiential dimensions of the learner are not visible in demographic data or skill assessments. They are visible in empathy research: the conversations where a learner describes what it felt like to complete the last training program, the observation session where a designer watches someone use a performance support tool under realistic time pressure and notices where the confusion surfaces, the focus group where participants talk about why they click through compliance training as fast as possible rather than engaging with it. This research is the LXD's distinctive analytical contribution to the IN phase, and it changes design decisions in ways that a purely cognitive needs analysis does not.

PROMPT: LXD Empathy Research Synthesis
I have conducted empathy research with [NUMBER] learners from [AUDIENCE] for [PROGRAM TOPIC]. Here are my research notes: Interview/observation notes: [PASTE OR SUMMARIZE] Key quotes that reveal emotional or experiential dimensions: [PASTE] Observed friction points (where confusion, frustration, or avoidance appeared): [DESCRIBE] Please synthesize this research to produce: 1. An emotional profile of this audience's relationship with this topic: - Default emotional state when approaching this content - Sources of professional identity investment or threat - Existing associations with how this topic has been trained before 2. A friction point map: the specific moments in a learning experience where this audience is most likely to disengage, click through, or experience confusion 3. A delight opportunity map: the specific moments where this audience is most likely to feel genuine recognition, competence, or connection 4. Design implications: 3-5 specific design decisions that this empathy data should change, compared to what a purely cognitive needs analysis would have recommended

LXD in Development: The Emotional Arc

Every learning experience has an emotional arc, whether it has been designed or not. In programs that have not been deliberately designed for emotional arc, the default arc is almost always the same: mild compliance anxiety at the start, low-grade boredom through the middle, relief at the end. This is the emotional arc of content delivery, and it is incompatible with learning.

The LXD's role in Development is to design the emotional arc deliberately: to map the emotional state the learner should be in at each stage of the program, and to design the specific elements, the scenario language, the activity structure, the feedback tone, the visual and interaction design, that create and sustain those states. This is not about manufacturing artificial enthusiasm. It is about creating the emotional conditions under which genuine learning is most likely to occur.

The Learning Experience Emotional Arc

The target emotional state at each stage of the learning experience.

Entry	Hook	Engagement	Challenge	Competence	Transfer
Skeptical or indifferent	*Curious or concerned*	*Invested in the problem*	*Productively stretched*	*Capable and confident*	*Ready to apply*

The arc above is a target, not a guarantee. It describes what the LXD is designing toward at each stage. The specific design elements that produce each emotional state depend on the audience, the topic, and the delivery format, and they require testing with real learners to confirm that the designed arc is the experienced arc. A scenario that the design team believes will produce curiosity may produce confusion in the audience. An activity the design team believes will produce productive stretch may produce discouragement. The pilot is the LXD's most important validation tool.

PROMPT: Emotional Arc Design Review
I need to design and then review the emotional arc for [PROGRAM/MODULE TITLE]. Target audience: [DESCRIBE — include emotional profile from empathy research] Program structure: [DESCRIBE MODULES, ACTIVITIES, SEQUENCE] Delivery format: [eLearning / ILT / blended]

Known audience emotional starting point: [FROM EMPATHY RESEARCH]

Part 1: Arc Design
For each stage of the program [list your stages], specify:
- The target emotional state at this stage
- The specific design element(s) responsible for producing it
- The risk: what might produce the wrong emotional state at this stage, and how the design mitigates it

Part 2: Arc Review (if content already drafted)
Please review the following content [PASTE CONTENT] and identify:
- Moments where the emotional arc is likely to hold
- Moments where it is likely to break (confusion, boredom, resistance, or the particular dread that compliance framing produces)
- Specific language or structural changes that would restore the arc at the identified break points

Note: Focus on the learner's emotional experience, not the content's logical organization. Those are different things.

Accessibility as Design Requirement

Accessibility is the area where the LXD orientation and a genuine commitment to inclusive design most clearly intersect, and it is the area most consistently treated as a compliance activity rather than a design principle. INDIE treats it as the latter. Accessibility is not a checklist applied at the end of development. It is a design constraint that shapes decisions from the first wireframe through the final QA review.

The case for accessibility-first design is not primarily legal, though the legal case is real. It is pedagogical: the design decisions that make learning experiences accessible to learners with disabilities, clear visual hierarchy, sufficient contrast, captioned audio, keyboard-navigable interfaces, unambiguous language, also make them better for learners without disabilities. A caption that was designed for a learner who is deaf is useful for a learner in a noisy open-plan office. Alt text written for a learner using a screen reader is useful for a learner whose image failed to load. Plain language designed for a learner with a cognitive disability is useful for a learner who is time-pressed and scanning rather than reading. Accessibility-first design is universal design, and universal design serves everyone.

Accessibility Criterion	INDIE Standard
Color and contrast	Minimum 4.5:1 contrast ratio for body text (WCAG 2.1 AA). Color is never the sole carrier of meaning — every color-coded distinction also has a text, pattern, or icon distinction.
Captions and transcripts	All video content has accurate, synchronized captions. All audio content has a transcript. Auto-generated captions are reviewed and corrected before publication; raw auto-captions do not meet INDIE standard.
Alt text	All meaningful images have descriptive alt text that conveys the information the image carries. Decorative images have null alt text. Alt text for complex diagrams describes the diagram's information, not its appearance.
Keyboard navigation	All interactive elements are reachable and operable via keyboard. Focus indicators are visible. Tab order is logical. No interaction requires a mouse as the sole input method.
Reading order	Screen reader reading order matches visual reading order. Heading hierarchy is logical (H1 → H2 → H3, no skipped levels). Lists are marked up as lists, not as visually indented text.
Timing	No content has a time limit that cannot be extended or disabled, except for activities where timing is essential to the learning objective and that exception is documented. Auto-advancing slides have a pause control.
Language	Primary language of content is declared. Technical jargon is defined at first use. Reading level is appropriate for the audience and has been verified, not assumed.
Motion and animation	Motion that serves no instructional purpose is eliminated. All motion can be paused or disabled. No content flashes more than 3 times per second (seizure risk threshold).
AI-generated content review	AI-generated text, images, and multimedia are specifically reviewed for accessibility before use. AI image generation tools often produce images with unreadable embedded text.

PROMPT: Accessibility Audit of Learning Content
I need to conduct an accessibility audit of the following learning content before it moves to production: Content type: [eLearning module / video / job aid / facilitator guide / etc.] Content: [PASTE TEXT CONTENT, DESCRIBE VISUAL/INTERACTIVE ELEMENTS] Authoring tool: [e.g., Articulate Storyline, Rise, Adobe Captivate, etc.] Target accessibility standard: [WCAG 2.1 AA / Section 508 / other] Please audit this content against the following criteria: 1. Color and contrast: Identify any color combinations that may fail 4.5:1 contrast ratio, or any meaning carried by color alone 2. Language clarity: Flag jargon, passive constructions, or reading level issues for the specified audience 3. Alt text gaps: Identify all images/icons described and evaluate whether the description conveys the informational content 4. Caption quality: If captions are present [PASTE TRANSCRIPT], identify accuracy errors, speaker identification gaps, and sound description omissions 5. Keyboard accessibility: Based on the interaction design described, identify elements that may require mouse-only interaction 6. Cognitive load: Flag sections where the density of information may create barriers for learners with cognitive disabilities For each issue found, specify: the location, the WCAG criterion affected, the severity (critical / major / minor), and a specific remediation.

When One Person Holds Both Roles

In many organizations, particularly smaller teams, nonprofits, and solo practitioners, there is no LXD. The instructional designer holds both roles simultaneously, either by design or by organizational default. This is not ideal, but it is common, and the INDIE framework accommodates it.

The practitioner who holds both orientations needs to be deliberate about when they are wearing which hat. The risk of holding both roles is not that you can't do the work; it is that the ID orientation, which tends toward systematic and analytical thinking, can crowd out the LXD orientation, which tends toward empathic and experiential thinking. Objectives get written before the learner's emotional relationship to the topic has been understood. Storyboards get drafted before the emotional arc has been designed. Accessibility gets reviewed at the end rather than built into the design from the start.

A practical discipline for the single practitioner: schedule explicit LXD time at the beginning of the IN phase (empathy research before objectives) and at the beginning of the Development phase (emotional arc design before storyboarding). These are not extra steps. They are the steps that prevent the more expensive rework that happens when an architecturally sound program reaches learners and produces the wrong emotional experience.

> **Role Callout: When You Hold Both Roles**
>
> A useful discipline for solo practitioners: after completing any significant design deliverable — a set of objectives, a curriculum map, a storyboard section, an assessment instrument, read it once as an ID (is this pedagogically sound?) and then read it again as a learner encountering it for the first time (how does this feel?). The questions are different, and they produce different observations. AI can support the second reading: a prompt asking, "What would a skeptical, time-pressed learner experience moving through this content?" reliably surfaces experiential issues that the ID reading misses, because the AI is not invested in the pedagogical architecture the ID has just spent an hour building.

Evan and Francis, Revisited

The product knowledge program took six weeks to design and build. The collaboration between Evan and Francis was, by both their accounts, occasionally frustrating and consistently productive.

Evan won the argument about establishing objectives first, with a modification: Francis's empathy research ran simultaneously, and the objectives were revised once the research came back. Two of the six initial objectives were rewritten, not because they were technically wrong, but because the empathy research revealed that the performance gap Evan had identified as primarily cognitive was, in the audience's actual experience, primarily a confidence gap. They knew the product features. They didn't trust themselves to discuss them fluently in a customer conversation. The objectives needed to address confidence and fluency, not knowledge recall.

Francis won the argument about understanding why the previous program had produced dread. The answer was specific: the previous program had been structured as a product feature inventory. It had asked learners to remember sixty-seven product attributes in eight modules. It had produced dread because it had been, genuinely, dreadful, not because the learners were resistant to learning, but because it had asked them to memorize rather than to practice. The new program was built around customer conversations, not product features. The features appeared in the scenarios, not in the content.

The pilot ran four weeks after the IN phase concluded. Level 1 scores were the highest either of them had seen for a product knowledge program in their respective careers. Several participants used the word "useful" in their open-ended feedback, which both Evan and Francis noted as significant: not "engaging," not "interesting," not the vocabulary of an experience that had been crafted to be enjoyable. Useful. The word that means: I know what to do now that I didn't know how to do before.

That is what happens when the two orientations work together. One designs the architecture. One designs the experience of moving through it. The learner gets both.

Chapter Summary

- The instructional designer and the learning experience designer are distinct orientations, each with real limitations when working alone. The ID produces rigorous architecture that can be experientially flat. The LXD produces compelling experiences that can be pedagogically hollow. Programs that achieve both require both orientations, in one practitioner or two.
- The ID's INDIE work is concentrated in IN-phase objective development and curriculum architecture, Development-phase Gagné design and assessment, and Enhancement-phase Kirkpatrick analysis. AI changes the economics of this work by compressing production time and freeing design capacity for the judgment and relational work that AI cannot do.
- The LXD's INDIE work is concentrated in IN-phase empathy research and emotionally-informed persona development, Development-phase emotional arc design and UX architecture, and accessibility-first design throughout. The learner's subjective experience of a program is not decorative. It is pedagogically consequential.
- The emotional arc of a learning experience should be designed deliberately, not left to default. The target arc moves from entry skepticism through curiosity, investment, productive challenge, and competence to transfer readiness. Each stage requires specific design decisions to sustain.

- Accessibility is a design requirement, not a compliance checklist. Accessibility-first design serves all learners, not only those with disabilities, and produces better learning experiences across every dimension of quality.
- For practitioners who hold both roles, the discipline is explicit hat-switching: schedule ID work and LXD work as distinct activities, and use AI to facilitate the LXD reading ("What does a skeptical learner experience here?) of work produced in ID mode.

Reflection Questions

For individual journaling or team discussion.

1. Which orientation, ID or LXD, more closely describes how you naturally approach a new design project? What is the characteristic limitation of your dominant orientation, and when has that limitation been visible in your work?
1. Think of a program you have encountered, as a designer, a facilitator, or a learner, that was architecturally sound but experientially flat, or experientially compelling but pedagogically hollow. Which orientation was missing, and what specifically would its presence have changed?
2. The chapter argues that the emotional arc of a learning experience is pedagogically consequential, not decorative. Do you agree? Have you seen evidence in your own work or in programs you have reviewed that the learner's emotional state influenced their learning outcomes? What would it take to make emotional arc design a standard part of your development process?
3. Apply the accessibility checklist in this chapter to a program you have recently developed or are currently developing. How many of the nine criteria can you confirm your program meets? For the criteria you cannot confirm, what does that reveal about where accessibility sits in your current design process?

Part Three · Chapter 11

The eLearning Developer and Multimedia Producer

Angela had built eLearning for nine years. She knew Storyline the way a carpenter knows a good saw, not just how it worked, but how it felt when it was working well, the particular satisfaction of a branching scenario that responded exactly the way the storyboard had intended, the animation timing that made an interaction feel fluid rather than mechanical. She had opinions about trigger sequencing. She had very strong opinions about auto-play audio.

The AI image generation tool had been in her workflow for six months. She had used it cautiously at first, then with growing confidence, then, after a specific incident she thought about more than she would have liked, with a great deal more care.

The incident was a compliance module for a healthcare client. She had used AI-generated images throughout: clinical settings, diverse medical staff, patient interactions. The images were clean, professional, appropriately diverse in the ways she had specified in her prompts. The client approved them. The module launched.

Three weeks later, a learner sent a screenshot to the L&D manager with a note attached. One of the images contained a medical monitor in the background displaying vital signs. The numbers on the monitor were physiologically impossible, a blood pressure reading that no living patient would produce. The image had passed through Angela's review, the client's SME review, and a two-stage quality check. Nobody had looked at the background closely enough to read the numbers. The module was pulled for revision. The client was understanding. Angela was not. "It looked right," she told her manager. "It looked exactly right. That's the problem."

The Roles That AI Changes Most Directly

If the previous chapter was about roles where AI changes the nature of the work, where it accelerates some tasks and liberates time for others, this chapter is about roles where AI changes the volume and character of production itself. The eLearning developer and the multimedia producer are the practitioners whose day-to-day output is most visibly and most immediately affected by the current generation of AI tools. They are also the practitioners for whom the quality risks of AI output are most concrete and most consequential.

These are production roles. Their output is not an analysis document or a curriculum map, it is a built artifact that learners encounter directly: a Storyline module, a video with narration, a set of scenarios with custom illustrations, an interactive branching simulation. The quality of that artifact is visible in ways that the quality of a needs analysis is not. A poorly reasoned objective stays in a design document. A poorly generated image ships in a module. The stakes of production quality are immediate and public.

This chapter addresses the two roles separately because their AI challenges, while related, are distinct. The eLearning developer's primary AI challenge is workflow integration: understanding which production tasks AI accelerates genuinely and which it accelerates in ways that introduce quality risks. The multimedia producer's primary AI challenge is media authenticity: understanding when AI-generated voice, image, and video serves the learner experience and when it undermines it. Both practitioners need a rigorous approach to quality control for AI-generated content that goes well beyond the surface-level review that traditional production workflows assumed.

The eLearning Developer in an AI-Augmented Workflow

The eLearning developer's traditional role sits at the intersection of instructional design and technical production. They take a storyboard, the design blueprint produced in the Development phase, and build it into a functional learning experience in an authoring tool like Articulate Storyline or Rise, Adobe Captivate, or Lectora. They implement the interactions, configure the branching logic, integrate the media assets, apply the visual design, and ensure that the technical output meets the specifications of the LMS environment it will be published to.

This is skilled work that requires both technical fluency and design judgment, the ability to recognize when a storyboard's intention cannot be executed as written, when an interaction needs to be redesigned for technical feasibility, when the visual design is creating cognitive load that the storyboard didn't anticipate. Good developers bring design sensibility to production work, and the distinction between a developer who executes and one who collaborates is visible in every module they build.

AI has entered this workflow at multiple points simultaneously. Content generation tools produce draft script text that reduces the volume of writing the developer needs to do before building. Image generation tools produce visual assets faster than stock libraries or custom illustration. Voice synthesis tools produce narration without recording schedules. Some authoring platforms have begun integrating AI assistance directly into their interfaces, suggesting interactions, generating quiz questions from content, proposing visual layouts. The production workflow that a developer learned three years ago looks materially different today, and the pace of that change is not slowing.

The AI-Augmented Production Workflow

The table below maps a typical eLearning development workflow against the specific AI contributions and the human judgment each stage still requires. It is organized around what actually happens in a production sprint, not around a theoretical framework, because developers think in production stages, not in instructional events.

Stage	AI handles	Human provides
Script Drafting	**AI handles** First draft of narration script from storyboard outline; plain language conversion of SME content; scenario dialogue; on-screen text alternatives to narration	**Human provides** Accuracy verification against source material; organizational voice calibration; cognitive load audit; removal of AI defaults (passive constructions, jargon, generic examples)

Rapid Proto-type	**AI handles** Initial slide structure from content outline; basic interaction scaffolding; placeholder visual layout; first-pass quiz items from learning objectives	**Human provides** Interaction design decisions requiring tool expertise; branching logic architecture; visual design judgment; determination of which interactions serve the learning and which are decorative
Visual Assets	**AI handles** Scenario character images; background environments; icon sets; simple diagrams; chart visualizations from data	**Human provides** Brand alignment review; demographic representation audit; technical accuracy verification of any depicted equipment or process; text-in-image legibility check; cultural appropriateness review
Narration & Audio	**AI handles** AI voice synthesis for procedural content; initial caption generation from script; audio leveling recommendations	**Human provides** Voice appropriateness decision for this content and audience; caption accuracy review and correction; emotional register evaluation; human re-recording decision for content requiring authenticity
Interactions & Branching	**AI handles** Scenario option text generation; feedback script drafts; branching outcome descriptions	**Human provides** Branching logic architecture requiring pedagogical judgment; consequence design that reflects real-world outcomes; feedback that teaches rather than evaluates; accessibility implementation
QA & Review	**AI handles** Consistency checks across slides; spelling and grammar review; link verification; reading level analysis	**Human provides** Pedagogical QA against learning objectives; cognitive load evaluation; media authenticity review; accessibility verification; brand compliance check; SCORM/xAPI technical testing

Rapid Prototyping: The Highest-Leverage AI Application

Of all the AI applications in the development workflow, rapid prototyping produces the most significant change in how eLearning development actually works. The traditional production model required a complete, approved storyboard before any building began because rebuilding was expensive, and the storyboard was the mechanism for catching design problems before they became production problems. Stakeholders reviewed storyboards because they could not see the module.

AI changes this. A developer who can generate a functional prototype from a content outline in a few hours changes the review conversation from "does this storyboard represent what you want?" to "does this prototype do what you need?" Those are different conversations. Stakeholders who struggle to visualize a storyboard often have immediate, specific reactions to a prototype. Design problems that survive multiple storyboard reviews surface immediately in a prototype because the interaction is concrete rather than described.

The discipline rapid prototyping requires is knowing what a prototype is for and holding that boundary with stakeholders. A prototype is a design validation tool. It is not a first draft of the final module. Stakeholders who conflate the two will request production-quality revisions on a prototype, consuming the time that the prototype was supposed to save. Managing that expectation, explicitly, before the prototype is shared, is the developer's responsibility and one of the most important facilitation skills the role requires.

PROMPT: Rapid Prototype Content Generation

I am building a rapid prototype for stakeholder review of [MODULE TITLE].
I need content that is representative enough to evaluate design decisions
but is clearly placeholder, not final.

Module topic: [DESCRIBE]
Target audience: [DESCRIBE]
Primary learning objective: [PASTE OBJECTIVE]
Prototype scope: [NUMBER] slides covering [DESCRIBE COVERAGE]
Key interaction to prototype: [DESCRIBE THE MAIN INTERACTION TYPE]

Please generate:
1. On-screen text for each prototype slide (headline + 2-3 body points)
   Written at the correct reading level for [AUDIENCE] but labeled
   clearly as DRAFT CONTENT throughout
2. Narration script draft for 3 key slides (the most complex content)
3. Scenario setup text for the prototype interaction:
   - Character and context (1 paragraph)
   - Decision prompt (1-2 sentences)

```
    - 3 response options with outcome descriptions
  4. Placeholder feedback for each response option

Flag any content where domain accuracy needs SME verification before
the prototype is shared with stakeholders outside the design team.

Note: This content will be used for design validation only.
All content requires expert review before production.
```

Role Callout: eLearning Developer

The most important habit change for developers in an AI-augmented workflow is building the stakeholder expectation conversation into the prototype delivery, not after it. Before sharing any AI-assisted prototype, send a brief framing note: "This prototype is for evaluating the interaction design and content structure. The narration is AI-generated placeholder text. The images are provisional. We are not yet in production. Your feedback on how the interaction works is what will move us forward." This framing prevents the prototype from being treated as a first draft of the final module, which is the most common and most expensive mistake in AI-assisted development workflows.

The Multimedia Producer and the Authenticity Problem

The multimedia producer's work, video production, audio recording, motion graphics, illustration, interactive media, sits at the intersection of creative craft and instructional purpose. Good multimedia production is invisible: the learner experiences the content, not the production. Bad multimedia production is very visible: the voice that feels mechanical, the image that looks wrong in a way the viewer can't immediately name, the video whose lighting makes the speaker look like they're being interrogated.

AI has given multimedia producers access to capabilities that were previously either expensive or impossible: voice synthesis that can produce narration in dozens of languages without recording sessions, image generation that can produce custom visual assets without illustration contracts, video generation that is rapidly approaching the quality threshold for some production contexts. These are genuine capabilities with genuine applications. They also introduce a set of quality and authenticity risks that the traditional multimedia production quality framework was not designed to catch.

The central challenge for the multimedia producer in an AI-augmented workflow is what might be called the authenticity problem: AI-generated media looks and sounds approximately right much more often than it looks and sounds exactly right. The gap between "approximately right" and "exactly right" is invisible at the speed of normal review, but it is visible to learners, often without their being able to articulate why the thing they're looking at feels slightly off. That feeling of slight wrongness is cognitively costly: it directs attention to the medium rather than the message, and attention directed at the medium is attention not available for learning.

Voice Synthesis: The Decision Framework

AI voice synthesis has reached a quality threshold at which, for many production contexts, it is genuinely indistinguishable from human narration on first listen. The question for multimedia producers is no longer "is the quality sufficient?" but "is this the right choice for this content and this audience?" Those are different questions, and the second one is harder.

The case for AI voice in learning is real: it eliminates recording scheduling friction, enables rapid content updates without re-recording costs, supports multilingual production at a scale that human narration cannot match economically, and produces consistent pacing and pronunciation across long-form content that even experienced voice talent sometimes does not. For procedural training, system walkthroughs, and high-volume content that requires frequent updating, AI voice is a genuine productivity advance without meaningful quality cost.

The case against AI voice is equally real, and it applies to specific contexts that many L&D programs include. Content that requires emotional authenticity, the safety story that is meant to land with weight, the leadership message that is supposed to feel genuine, the difficult conversation scenario that depends on emotional texture for its instructional power, loses something meaningful when the voice producing it is demonstrably synthetic. The loss is not about audio quality. It is about the signal that a human voice sends: someone cared enough about this content to speak it. AI voice sends a different signal, and in contexts where the human signal matters, the AI voice undermines the content it is meant to carry.

AI Voice Decision Framework

A guide for multimedia producers making the AI voice decision for specific content and audience contexts.

Context	AI Voice: Appropriate	AI Voice: Undermines Trust
Content type	Procedural instruction, system walkthroughs, process explanations, reference content, technical documentation	Leadership messages, personal testimonials, coaching conversations, sensitive topics (mental health, safety incidents, ethical dilemmas), content requiring emotional authenticity
Audience relationship to narrator	Audience has no expectation of a real person behind the voice; the narrator is understood as a neutral guide	Audience expects a real person, an executive message, a subject matter expert, a peer story, a human moment of the kind that compliance training often inappropriately attempts
Content frequency and volume	High-volume, frequently updated content where re-recording human voice would be cost-prohibitive, software update walkthroughs, regulatory updates, product catalog changes	Signature programs, flagship leadership development experiences, or any content where the organization's brand voice is a meaningful part of the learner experience
Cultural and linguistic context	Global content where consistent neutral delivery is preferred over accented human voice that may disadvantage non-native speakers	Content where regional authenticity matters, local leadership voices, culturally specific examples, community-facing programs where the voice is part of the message of inclusion
Learner population trust level	Audiences with high tolerance for digital inter-	Audiences with low digital trust, frontline workers who have experienced technology as distancing

	faces and no strong expectation of human presence in training delivery	rather than connecting, or populations where the human-to-human signal is part of what the organization is trying to communicate

The AI Voice Quality Check Most Producers Miss

AI voice synthesis produces natural-sounding speech at the sentence level but frequently fails at the paragraph level in three specific ways. First, pacing: AI narrators do not breathe, and the absence of breath pauses creates a subtle sense of unreality in longer passages that audiences register as "something is slightly off" without identifying the cause. Second, emphasis: AI voice applies emphasis based on linguistic patterns in training data, which frequently misaligns with the instructional emphasis the content requires. A sentence like "You must verify the patient's identity before administering medication" may receive AI emphasis on "patient's identity" rather than "before," a subtle but consequential difference. Third, emotional register: AI voice is trained to sound professional and pleasant, which is appropriate for most content and actively wrong for content that requires gravity, urgency, or genuine warmth. Review AI voice at the paragraph level and listen for all three. Correct in the script (pace markers, emphasis markup) or re-record with human talent when correction is insufficient.

PROMPT: AI Voice Script Optimization

I have the following narration script for an AI voice synthesis tool.
I need to optimize it for natural AI delivery before synthesis.

Script: [PASTE NARRATION SCRIPT]
Voice style target: [e.g., warm and collegial / authoritative and clear / conversational and direct]
Content type: [PROCEDURAL / EXPLANATORY / SCENARIO-BASED / OTHER]
Known AI voice tool: [e.g., ElevenLabs, Murf, Descript, Speechify]

Please review and optimize this script for AI voice delivery:

1. Sentence length: Break sentences longer than 20 words into two. Long sentences produce AI pacing problems.

2. Emphasis markers: Insert [EMPHASIS] before any word or phrase that carries instructional weight. Flag where AI default emphasis may conflict with instructional intent.

3. Pause markers: Insert [PAUSE] at points where a human narrator would breathe or where the listener needs processing time. Minimum: one pause per paragraph.

4. Emotional register flags: Mark any sentence where the content requires an emotional register (gravity, warmth, urgency) that AI voice is unlikely to produce. Recommend human re-recording for those sections.

5. Jargon and acronyms: Flag any term that AI voice may mispronounce and provide a phonetic alternative.

Return the optimized script with all markers in place.

Image Generation: What Goes Wrong and How to Catch It

Angela's incident, the blood pressure monitor with impossible numbers, is one of a specific class of AI image generation failure that has a common cause: AI image generators are trained to produce images that look visually coherent, not images that are factually accurate. The difference is consequential in learning contexts where visual accuracy matters. A medical monitor that looks like a medical monitor but displays impossible values is worse than no monitor at all; it teaches learners what medical monitors look like incorrectly.

This failure class extends well beyond healthcare. Equipment in manufacturing training images that shows incorrect safety configurations. Legal documents in compliance training images that contain placeholder text that looks like real legal language. Software interfaces in technology training that show menu structures that don't match any real product. Architectural details in facilities training that violate code requirements. In every case, the image looks approximately right, passes a surface-level review, and contains a specific technical error that a subject matter expert would catch instantly and a learning professional without domain expertise might not.

The systematic response to this failure class is not to stop using AI image generation. It is to build domain-specific accuracy review into the production workflow as a standard step. Every image that depicts specialized equipment, processes, documents, or environments requires review by someone with domain expertise, not for visual quality, but for technical accuracy at the detail level. This is a step that did not exist in traditional stock photography review because stock photography was produced by photographers who had access to real equipment and real environments. AI-generated imagery was not.

The Five Failure Modes of AI-Generated Learning Images

1. Embedded text errors: AI generates text within images (on signs, monitors, documents, interfaces) that is visually plausible but factually wrong, spelled incorrectly, or linguistically incoherent. Check every image that contains any text at pixel level, not at thumbnail resolution. **2. Anatomical and physiological inaccuracies:** AI medical, healthcare, and human-performance images frequently contain subtle anatomical errors, incorrect joint positions, implausible grip configurations, body proportions inconsistent with realistic human movement under load. **3. Equipment configuration errors:** AI-generated equipment, machinery, and technology images show configurations that are visually coherent but operationally impossible or safety-violating. **4. Demographic defaults:** Without explicit prompt specification, AI image generation defaults to a narrow demographic range. Prompts specifying "a professional in an office" produce images that are not representative of the learner population or the broader workforce. **5.** Cultural context errors: AI images depicting workplace or social contexts frequently contain cultural markers, gestures, spatial arrangements, symbolic objects, that are accurate for one cultural context and incorrect or offensive in another. Global programs require cultural review of all AI-generated imagery by someone with regional expertise.

PROMPT: AI Image Generation for Learning Contexts

I need to generate images for a learning module on [TOPIC] for [AUDIENCE]. This prompt is designed to minimize the common failure modes of AI image generation in instructional contexts.

Image needed: [DESCRIBE WHAT THE IMAGE SHOULD SHOW]
Purpose in the module: [WHAT INSTRUCTIONAL FUNCTION DOES THIS IMAGE SERVE]
Domain accuracy requirements: [LIST ANY TECHNICAL ELEMENTS THAT MUST BE
ACCURATE, equipment, documents, processes, interfaces]

Please generate a detailed image prompt that:

1. Specifies the demographic composition explicitly:
   [DESCRIBE DESIRED CHARACTERS, age, gender expression, apparent ethnicity, ability status as appropriate to the context]

2. Specifies what should NOT appear to avoid failure modes:
   - No visible text on any surface unless explicitly specified and provided verbatim in this prompt
   - No equipment with visible controls, displays, or readouts unless

```
    the required values are specified exactly
  - No gestures that have different meanings across cultural contexts

3. Specifies the visual style:
  [PHOTOREALISTIC / ILLUSTRATED / SEMI-REALISTIC] appropriate to
  the brand and audience

4. Identifies the elements that will require domain accuracy review
  before this image can be approved for production use

After generating the prompt, list the specific technical elements in
this image that should be reviewed by a domain SME before approval.
```

Quality Control for AI-Generated Media: The Producer's Framework

Traditional multimedia QA frameworks were built for human-produced content, where the primary quality risks were technical: compression artifacts, audio sync errors, color calibration inconsistencies, resolution mismatches. These risks still exist in AI-augmented production. What AI adds is a category of quality risk that traditional frameworks do not address: the content that looks technically correct but is substantively wrong.

The media QA checklist below reorganizes the review process around both categories. It is designed to be applied asset-by-asset during production, not only at the final review stage because catching an AI-generated error during asset production is significantly less expensive than catching it during a stakeholder review or, as Angela experienced, after launch.

Quality Criterion	What to Check
Script accuracy	Every factual claim in the narration script has been verified against an authoritative source. AI-generated script content has been reviewed by a domain SME for any specialized content area. No AI-generated citations, statistics, or regulatory references appear without independent verification.

Voice appropriateness	The AI/human voice decision has been made explicitly against the decision framework, not by default. If AI voice is used, the script has been optimized with pace markers, emphasis flags, and emotional register review. Sections requiring human re-recording have been identified.
Caption accuracy	All captions have been reviewed against the script word-for-word, not speed-read. Speaker changes are identified. Significant non-speech audio (alarms, notifications, ambient sound with instructional relevance) is described. Auto-generated captions have not been published without correction.
Image technical accuracy	All images depicting equipment, documents, interfaces, or processes have been reviewed by a domain SME for technical accuracy at the detail level. Text visible in any image has been verified for accuracy and correct spelling. No placeholder text is present in final assets.
Image text legibility	All text visible in images is legible at the display resolution of the target device. Text in images meets WCAG contrast requirements independently of the surrounding slide design.
Audio-visual sync	Narration timing matches on-screen content timing. Animation triggers do not conflict with narration pacing. Learner-controlled interactions have sufficient pause for interaction before narration resumes.
Brand consistency	AI-generated visual assets are consistent with organizational brand guidelines. Color palette, typography, and illustration style match approved assets. AI-generated imagery has not introduced visual styles inconsistent with the brand.
Interaction accuracy	Branching logic produces the outcomes described in the storyboard. Feedback text is accurate and pedagogically appropriate. No dead ends or loop errors in navigation. All variables and triggers function as intended.
Technical compliance	SCORM/xAPI metadata is accurate. Completion and scoring logic matches LMS requirements. Module behaves correctly in the target browser and device environments. No broken links or missing assets.

The Authoring Tool Workflow: Practical Integration

The specific integration of AI tools with authoring platforms like Articulate Storyline, Rise 360, Adobe Captivate, and Lectora changes with each software update, too quickly for this book to serve as a current technical reference. What is more durable is the integration philosophy: a set of principles for how AI tools should fit into the authoring workflow that will apply regardless of which specific tools are in use when you read this.

- **AI before authoring, not during it.** The most effective use of AI in the authoring workflow is generating and refining content before it enters the authoring tool, not within it. Script drafting, scenario writing, quiz question generation, and image prompt development should happen in an AI environment where iteration is fast and revision is frictionless. Revising content inside an authoring tool is slower and more structurally fragile. Build clean, reviewed, approved content and then build it in the tool, not both simultaneously.
- **Prototype in AI, produce in the tool.** Rapid prototypes that use AI-generated placeholder content to validate design decisions can be built quickly in the authoring tool as lightweight structural shells. When the design is validated, replace placeholder content with reviewed, approved production content. Treat these as two separate build phases with two separate review gates, not a single continuous build process.
- **Version control everything AI touches.** AI-generated content that enters a production workflow creates version control challenges that human-written content does not. An AI tool will produce slightly different output for the same prompt on different occasions, meaning that "regenerate" is not the same as "restore." Every AI-generated asset that enters production should be saved in its reviewed and approved form before it is used, and the prompt that generated it should be documented for future reference.
- **Never publish auto-generated captions.** This point deserves its own line because it is violated constantly. Auto-generated captions from AI transcription tools contain errors that are invisible

at normal reading speed and significant to learners who depend on them, particularly technical vocabulary, proper nouns, and domain-specific terminology that transcription models consistently misrepresent. Every caption file requires a word-for-word review against the source script before publication. There is no workaround for this. "Good enough" captions are not good enough.

- **Build the SME accuracy review into the timeline, not after it.** The domain accuracy review that AI-generated content requires is a production step that should appear in the project timeline as a named milestone, not as an informal check that happens when someone thinks of it. Identify the SME reviewer at the start of the project, specify what they are reviewing and what they are not reviewing, allocate time in the schedule, and build their feedback into the revision cycle. Accuracy review that is treated as optional will be deferred under deadline pressure and will surface in the Enhancement data.

> **Role Callout: Multimedia Producer**
>
> The professional skill that has become most valuable for multimedia producers in an AI-augmented environment is not fluency with AI generation tools, those are learnable quickly and change rapidly. It is the critical eye that can identify what AI-generated media cannot do: the emotional register that a voice actor brings to a safety story, the specific cultural authenticity that a location-produced video carries, the technical accuracy that only a practitioner with domain expertise can verify. Producers who develop and articulate this critical eye become the quality gatekeepers that AI-augmented workflows require. Those who do not become the people who launch modules with impossible blood pressure readings.

Angela, Revisited

The blood pressure module was revised within a week. Angela rebuilt the review process before the next project started. She added a step she called "pixel-level read" to her production QA: for every image that contained equipment, documents, or environments with visible details, she zoomed in to 200 percent and read every visible element. She built a domain accuracy review step into her project timelines as a named milestone, a thirty-minute video call with the relevant SME, scheduled before final production, with a specific checklist of the technical elements she needed verified.

She also changed how she talked about AI-generated images with her clients. Not defensively, she did not stop using them, but clearly. "These are AI-generated assets. They are visually accurate at the level we can review in this meeting. Before we publish, I need a thirty-minute session with your clinical lead to verify the technical elements. That's a step in our process now." The clinical lead, when she heard this, said: "Thank you. I've been worried about this."

The next program Angela built for the healthcare client had AI-generated images throughout. The clinical lead reviewed every image with visible medical equipment. She found two additional issues that Angela's pixel-level read had missed, both subtle, both correctable, both the kind of thing that would have been found, eventually, by a learner who knew what they were looking at.

The module launched without incident. The SME review step added four hours to the project timeline. The alternative, a post-launch pull and revision, had cost three weeks the first time.

Angela had not stopped trusting AI-generated imagery. She had learned exactly where it needed to be verified, and she had built the verification in. That is what quality control looks like in an AI-augmented production workflow: not less use of AI, but clearer-eyed use, with the human expertise applied precisely where it is needed most.

Chapter Summary

- The eLearning developer and multimedia producer are the roles most immediately affected by AI production tools, and the roles for whom the quality risks of AI output are most concrete and consequential, because their output is a published artifact learners encounter directly.
- AI-augmented development workflow maps onto all production stages: script drafting, prototyping, visual asset generation, narration, interactions, and QA. In each stage, AI handles volume and speed while human expertise provides accuracy verification, design judgment, and quality gatekeeping.
- Rapid prototyping is the highest-leverage AI application for developers: it moves stakeholder review from storyboard description to functional experience, surfacing design problems earlier and more concretely. The discipline required is managing the prototype-as-design-tool boundary with stakeholders.
- The AI voice decision is not a quality question but an appropriateness question: AI voice is appropriate for procedural and high-volume content where human presence is not part of the instructional intent. It undermines trust in content requiring emotional authenticity, leadership presence, or human-to-human signal.
- AI image generation produces five specific failure modes in learning contexts: embedded text errors, anatomical inaccuracies, equipment configuration errors, demographic defaults, and cultural context errors. All five require systematic review processes that traditional stock photography QA did not include.
- The media QA framework must expand beyond technical quality (sync, resolution, compression) to include substantive accuracy review: content verified against authoritative sources, technical elements reviewed by domain SMEs, and captions verified word-for-word against source scripts.
- Integration principles: AI before authoring not during it; prototype in AI and produce in the tool; version control everything AI touches; never publish auto-generated captions; build SME accuracy review into the timeline as a named milestone.

Reflection Questions

For individual journaling or team discussion.

1. Apply the workflow table in this chapter to a production project you are currently working on or have recently completed. Which stages are you using AI for, and which are you not? For the stages where you are not using AI, what is the barrier, tool availability, trust, workflow integration, or something else?
2. Using the AI voice decision framework, review the last three modules you have produced that included narration. For each, was the human/AI voice choice made explicitly against a framework, or by default? If by default, would the framework have produced the same decision or a different one?
3. Angela's incident, the impossible blood pressure reading that passed multiple reviews, illustrates how AI-generated accuracy errors are invisible at normal review speed. Think of the AI-generated content in your most recent production project. What domain-specific technical elements did it contain that required accuracy review by someone with domain expertise? Did that review happen? If not, what would it have cost to build it in?
4. The chapter argues that never publishing auto-generated captions without correction is non-negotiable. Be honest: have you published auto-generated captions in the past without a word-for-word review? What would it take to make caption review a standard step in your production workflow rather than an aspirational one?
5. The multimedia producer role callout argues that the critical eye that identifies what AI-generated media cannot do is the most valuable skill for producers in an AI-augmented environment. Based on your experience, what does AI-generated media consistently fail to produce that you have had to supply through human production or human judgment? How would you articulate that capability to a hiring manager or a colleague?

Part Three · Chapter 12

The Facilitator and Trainer

Jerome had been facilitating training for sixteen years. He had delivered leadership programs and compliance sessions, technical workshops and culture initiatives, half-day seminars for twelve people and two-day programs for two hundred. He had facilitated rooms where people cried and rooms where people walked out and rooms where nothing happened at all because the organization had sent the wrong people on the wrong day for the wrong reasons. He knew what it felt like when a room was alive and what it felt like when a room was dead, and he knew, usually within the first eight minutes, which one he was in.

He had always prepared carefully. Not obsessively: he had learned early that over-scripted facilitation was its own kind of failure, the facilitator so committed to the plan that the room ceased to exist as anything other than an obstacle to executing it. But he prepared with intention: the audience research, the three possible ways the opening activity might go and what he would do in each case, the question he would ask if the conversation went somewhere important, the two or three things he absolutely needed to land no matter what else happened.

When AI tools arrived in his workflow, he was initially uninterested. Facilitation, as he understood it, was a human act performed in a human room. What could a language model contribute to the moment when someone said something vulnerable and the whole group leaned forward?

The answer, he eventually concluded, was: nothing. But that was not the right question. The right question was: what could AI contribute to the hours before that moment, so that when it arrived, he was as prepared as it was possible to be? The answer to that question was: quite a lot.

What Facilitation Actually Is

Before addressing how AI changes the facilitator's work, it is worth being precise about what that work is because facilitation is one of the most consistently misunderstood roles in the L&D field, both by people outside it and occasionally by practitioners within it.

Facilitation is not the delivery of content. A presenter delivers content. A facilitator creates the conditions under which a group can think together, question assumptions, apply new ideas to real problems, and arrive at understanding through their own cognitive and social engagement rather than through passive receipt. The distinction is not semantic. A presenter's success is measured by how clearly they communicate. A facilitator's success is measured by what the group produces.

This distinction shapes everything about how facilitation should be prepared and evaluated. A facilitator who prepares primarily by mastering the content is preparing for the wrong job. A facilitator who prepares by mastering the content, the audience, the process design, and the range of ways that process can unfold is preparing for the actual job. The content knowledge is a prerequisite. The facilitation skills, reading a room, managing group dynamics, asking the question that unlocks the conversation, holding silence without filling it, redirecting constructively when a discussion goes somewhere that doesn't serve the learning, are the job.

AI's role in the facilitation context is almost entirely pre-session: helping the facilitator prepare more thoroughly and think through more scenarios than time would otherwise allow. The facilitation itself remains a human act, irreducibly dependent on the facilitator's presence, judgment, and relationship with the room.

Theoretical Grounding: Vygotsky and the Social Construction of Knowledge

Facilitated learning draws most directly on Vygotsky's social constructivism: the idea that knowledge is not transferred from teacher to learner but constructed by learners in social interaction. The facilitator's role, on this view, is not to transmit what they know but to create the conditions under which the group can construct understanding together, structuring the interactions that produce cognitive engagement, naming the patterns that emerge in discussion, mediating between perspectives, and holding the space in which genuine thinking can occur. Lave and

Wenger's concept of legitimate peripheral participation extends this: learners in a facilitated program are being inducted into a community of practice, not just acquiring discrete knowledge. The facilitator manages that induction, making the community's norms visible, connecting new knowledge to the community's existing practice, and helping learners see themselves as capable members of that community. These are acts of presence and relationship that AI cannot perform.

AI in the Facilitator's Workflow: Before, During, and After

The facilitator's workflow has three distinct phases: preparation (everything before the session), delivery (the session itself), and synthesis (what happens immediately after). AI has meaningful applications in the first and third. In the second, the delivery itself, AI's role is minimal, and the attempt to use it actively during facilitation is more likely to undermine presence than support it.

Before: Preparation as Practice

The most common facilitation failure is not incompetence in the room. It is under-preparation before it. Facilitators who know the content but have not thought through the process, who have not anticipated the moments where the conversation might stall, the participants who might dominate or withdraw, the question that reveals the group hasn't understood what they thought they understood, arrive in the room with insufficient cognitive bandwidth for the actual work of facilitation. They are managing the content while the room manages itself, which is not the arrangement that produces good outcomes.

AI cannot replace the experience and judgment that makes a skilled facilitator effective. It can dramatically increase the quality of preparation by expanding the range of scenarios a facilitator has thought through before walking in. A facilitator who has worked through six possible ways the opening discussion might unfold is more likely to respond well to the actual seventh way than one who has only thought through one.

Preparation Area	What to Do and Why It Matters
Audience intelligence	Before any session, research the specific participants: their roles, their likely prior experience with this topic, any recent organizational events that might shape their mood or resistance level. AI can synthesize available information and generate a participant profile. The facilitator's judgment about what matters is irreplaceable, but AI can surface information the facilitator might not have sought.
Session purpose clarity	State the session's purpose in one sentence before preparing anything else: not the learning objectives (that's the designer's frame) but the facilitation purpose: what should the group have that they didn't have when they walked in? Clarity of purpose is the most reliable predictor of facilitation quality. AI can help test whether a stated purpose is genuinely clear by attempting to derive the session design from it.
Process design review	Work through the session design as a sequence of group experiences, not a schedule of activities. For each segment: what is the group doing cognitively? what are they producing? how does it connect to what came before and what comes after? AI can review a session plan for process logic and flag segments where the connection between activity and purpose is unclear.
Anticipating the hard moments	Identify the two or three moments in the session where facilitation skill matters most: the activity that might fall flat, the discussion question that might go nowhere or everywhere, the transition that loses the room. For each, think through three possible responses. AI is excellent for generating these alternatives; it produces scenarios the facilitator might not have considered, at a range of difficulty levels.
Opening and closing design	The opening and closing of a session have disproportionate impact on what participants remember and carry back to their work. Prepare both with more care than the middle sections receive. AI can generate multiple opening options and closing synthesis approaches for a given topic and audience, allowing the facilitator to select the approach that best fits their read of the group.
Content boundary mapping	A facilitator who does not know where their content expertise ends is a facilitator who will be embarrassed in the room. Map the questions this session is

	likely to generate that fall outside the facilitator's confident knowledge. For each, prepare either a genuine answer or a graceful redirect. AI can generate the questions the audience is likely to ask, including the ones that fall outside the planned content.

PROMPT: Session Preparation: Anticipating the Room
I am facilitating [SESSION TITLE] for [AUDIENCE DESCRIPTION]. Session length: [DURATION]. Format: [IN-PERSON / VIRTUAL / HYBRID]. Here is the session plan: [PASTE OR DESCRIBE SESSION PLAN] Please help me prepare by generating: 1. Participant profile: Based on [ROLE/DEPARTMENT/CONTEXT], what are the most likely: - Prior experiences with this topic (positive and negative) - Skepticisms or resistances I should anticipate - Questions that are likely to come up that go beyond the planned content - Dynamics between participant groups if multiple departments or levels are present 2. Hard moment anticipation: For each major segment of this session, identify the most likely facilitation challenge: - Activity that might fall flat and why - Discussion that might go nowhere (silence) or everywhere (too much energy in the wrong direction) - Transition that might lose the room For each challenge, provide three possible responses: one that redirects, one that names the dynamic, one that adapts the plan. 3. Opening alternatives: Generate 3 different ways to open this session that each create a different emotional register: - Option A: Creates curiosity and safety - Option B: Creates urgency and relevance - Option C: Activates prior experience and honors expertise 4. Closing synthesis: Generate 2 approaches to closing this session that help participants name what they will do differently as a result of what they experienced today.

Before: Facilitating Content You Didn't Design

The most important facilitation preparation challenge that AI has created is not new in kind but new in scale: many facilitators now deliver programs they had no hand in designing, built with AI assistance by a designer who may be in a different city, who had a limited window to prepare the facilitation guide, and who made design decisions the facilitator was not present to understand.

This is not a theoretical concern. As AI accelerates the production side of the development workflow, the speed differential between building a program and preparing a facilitator to deliver it has widened. A designer who can produce a facilitation guide in two days with AI assistance has still produced a guide that a facilitator needs meaningful time to internalize. The guide arrived faster. The preparation time needed did not decrease.

For facilitators in this position, delivering programs they didn't design, with limited preparation time, carrying a guide produced by someone who may be unavailable for questions, AI provides a specific and high-value preparation support: it can help the facilitator interrogate the guide's design logic, anticipate the moments the guide doesn't address, and build a mental map of the session that the facilitator owns, rather than following someone else's map one step at a time.

PROMPT: Internalizing a Facilitation Guide You Didn't Design
I am facilitating [SESSION TITLE] using a guide I did not design. I have [TIMEFRAME] to prepare. Here is the facilitation guide: [PASTE FACILITATION GUIDE] Please help me build a mental model of this session by: 1. Design logic summary: In 3-5 sentences, explain the pedagogical logic of this session design. What is it trying to accomplish, and why is it structured this way? (Help me understand the "why" behind the sequence, not just the "what.") 2. Critical moments: Identify the 3 moments in this guide where the facilitation will most determine the outcome. What is required of the facilitator at each moment?

3. Guide gaps: What does this guide not address that I should be prepared for anyway? (Common audience questions, likely resistance points, transitions that need more than the guide provides)
4. Flex points: Where in this session is there flexibility to adapt if the room is going differently than planned? And where are the non-negotiables — the elements I must not cut regardless of time pressure?
5. My 3 anchor questions: Generate 3 questions I can use at any point in this session to bring the group back to the central purpose if the discussion drifts.

During: What the Facilitator Carries Into the Room

The session itself is where AI recedes and the facilitator's preparation returns on investment. The hours spent with AI generating scenarios, anticipating hard moments, and building a mental map of the session exist for one purpose: to give the facilitator enough cognitive space in the room to be fully present to what is actually happening.

Presence is not a soft skill. It is the core technical capacity of skilled facilitation: the ability to track the group's cognitive and emotional state in real time, to notice when a participant's body language contradicts their verbal agreement, to sense when the energy in the room has shifted from engagement to performance, to hold the space when a genuine moment of learning is occurring without rushing to fill it with more content. These capacities require cognitive bandwidth, and cognitive bandwidth is consumed by uncertainty. A facilitator who is uncertain about the process, uncertain about the content, uncertain about what comes next has no bandwidth for presence. A facilitator who is thoroughly prepared and confident in their map of the session can set that map down and be in the room.

There is a specific question about AI use during virtual facilitation that is worth addressing directly: can AI tools be used during a live session to support facilitation in real time? The answer is yes, in limited ways, with significant caveats. Virtual breakout rooms allow a brief window for a facilitator to query an AI tool for a discussion prompt or a content clarification. The risk is that the cognitive bandwidth spent interacting with an AI tool during a session is bandwidth diverted from the room. A facilitator who is reading an AI-generated response while participants are returning from a breakout is a facilitator who is not watching how participants look when they come back, which is often where the most important information about the session's progress lives.

After: Synthesis and Transfer Support

The work that happens in the twenty-four to forty-eight hours immediately following a facilitated session has an impact on learning outcomes that most facilitation practice systematically underinvests in. Ebbinghaus's forgetting curve begins the moment the session ends. What participants consolidate in the hours after a session determines a significant proportion of what they retain and transfer. The facilitator is uniquely positioned to support that consolidation; they were in the room, they know what the group produced and what it struggled with, and they can provide synthesis that is specific to this group rather than generic to the topic.

AI accelerates the post-session work that facilitators know they should do and often don't, because the session itself is exhausting and the next session is approaching. A facilitator who can produce a session synthesis document in forty minutes rather than two hours is a facilitator who will actually produce it, which means participants receive it while the experience is still fresh, rather than three days later when they've moved on.

PROMPT: Post-Session Synthesis Document
I have just facilitated [SESSION TITLE] with [NUMBER] participants. Here are my notes from the session: Key themes that emerged in discussion: [DESCRIBE] Moments of insight or breakthrough: [DESCRIBE SPECIFIC MOMENTS] Questions that were raised but not fully resolved: [LIST] Activities that generated strong engagement: [DESCRIBE] Activities that fell flat or needed adjustment: [DESCRIBE] Commitments or intended actions participants mentioned: [LIST IF CAPTURED]

Please produce a post-session synthesis document for participants that:

1. Opens with a 2-sentence framing of what this group accomplished together (specific to THIS group, not generic to the topic)

2. Captures the 3-5 most significant insights that emerged from the group's discussion (attributed to the group, not to the facilitator or the content)

3. Names the questions that the group surfaced but didn't fully resolve, framed as productive tensions worth continued reflection

4. Connects the session's content to one concrete application opportunity for this specific audience in their specific context

5. Closes with one question for individual reflection in the next 48 hours

Tone: Warm, specific, collegial. Speaks to what THIS group experienced, not what people generally experience in sessions on this topic.
Length: One page maximum.

PROMPT: Facilitator Debrief for Enhancement Data

I need to capture facilitation observations for the Enhancement phase of [PROGRAM TITLE]. I facilitated [NUMBER] sessions this [WEEK/MONTH].

My observations across sessions:
Activities that consistently generated strong engagement: [DESCRIBE]
Activities that consistently fell flat: [DESCRIBE]
Content sections where confusion appeared repeatedly: [DESCRIBE]
Questions participants raised that the content didn't address: [LIST]
Moments where the facilitation guide was insufficient: [DESCRIBE]
Differences between participant groups: [DESCRIBE IF RELEVANT]

Please organize these observations into a structured facilitation debrief for the design team that:

1. Separates design issues (content, structure, activities) from delivery issues (facilitation guide, pacing, materials) from context issues (room, timing, audience preparation)

2. Prioritizes issues by frequency and impact

3. For each high-priority design issue, proposes a specific revision and explains the facilitation evidence that supports it

4. Identifies any pattern suggesting the program objectives may not be matching participant needs as experienced in the room

Format this for a designer who was not in the room, who needs to understand what the facilitation experience revealed about the design.

The Hard Rooms: Facilitation Challenges That Preparation Serves

No facilitation preparation framework is complete without an honest account of the rooms that preparation doesn't fully solve, the situations where the facilitator's live judgment and interpersonal skill are tested in ways that no amount of AI-assisted scenario generation fully replicates. What AI can do is help the facilitator think through these situations before they happen, so that when they happen, the response comes from preparation rather than improvisation.

The scenarios below are the ones experienced facilitators cite most often as the moments where they have felt most tested. For each, the card names the signal that the situation is occurring, a response that has repeatedly proven effective, and the response to avoid, which is often the one that feels most natural in the moment.

The Expert Who Knows More Than You

Signal: *A participant visibly signals domain expertise greater than the facilitator's, through a question the facilitator can't fully answer, a correction of content, or a posture of polite tolerance.*

Response: Name it directly and redirect to the group's resource: "You know this area better than I do, which is useful for everyone here. Can you say more about how that plays out in your context?" The expert becomes a resource rather than a threat. The facilitator's role shifts from content authority to process guide, which was the correct role all along.

Avoid: Defend or deflect. Both are visible, both reduce credibility, and neither serves the learning. A facilitator who doesn't know something and says so clearly is more credible, not less, than one who obscures the gap.

The Dominant Voice That Shuts Others Down

Signal: *One participant speaks significantly more than others, others have stopped attempting to contribute, and the group's energy has become an audience watching a performance.*

Response: Use structure to redistribute voice without targeting the individual: "I want to make sure we hear from everyone on this before we move on. Let's go around, one sentence from each person on what you'd add." Alternatively, use small group work to create space where the dominant voice is only one of three or four. The goal is structural, not corrective.

Avoid: Address the individual directly in the room. It embarrasses them, creates defensiveness, and makes the group uncomfortable on their behalf. Save direct feedback for a break or after the session.

The Resistant Room

Signal: *The group is compliant but disengaged, responding to questions with minimal elaboration, completing activities without investment, producing the form of participation without the substance.*

Response: Name the dynamic without blame: "I'm sensing this might not be landing the way I hoped. Can I ask, what would make this more useful for you today?" This requires courage, but it often breaks the dynamic entirely. A group that has been asked what it needs is a group that has been invited into collaboration rather than subjected to delivery.

Avoid: Power through. A facilitator who responds to a resistant room by trying harder, more energy, more enthusiasm, more content, usually produces more resistance. The room is telling you something. Listen to it.

The Unexpected Emotional Disclosure

Signal: *A participant shares something personal, difficult, or emotionally significant in response to a content prompt. The room goes quiet. The facilitator is not a therapist and this is not a therapy session, but something real has happened and needs to be acknowledged.*

Response: Acknowledge the disclosure with genuine care, without amplifying or analyzing it: "Thank you for sharing that. It's clearly connected to why this matters." Then offer the participant control: "Would you like to say more, or shall we hold that and keep moving?" Let them choose. The group's humanity has been activated; honor it briefly and continue with respect.

Avoid: Either ignore the disclosure (which signals that the facilitator didn't hear it, or didn't think it mattered) or over-engage with it (which shifts the session into territory the facilitator is not equipped to navigate and may not serve the participant or the group). Both of these are common. Neither is kind.

The Content Question You Can't Answer

Signal: *A participant asks a specific, substantive question that falls outside the facilitator's confident knowledge. The group is watching to see what happens.*

Response: "I don't know the answer to that with confidence, and I'd rather find out than guess. I'll get you an answer after the session. Does anyone here have experience with this?" This response demonstrates intellectual honesty, redirects to the group's collective knowledge, and models the learning behavior the session is trying to develop. It is not a facilitation failure. It is a facilitation choice.

Avoid: Attempt to answer from insufficient knowledge. Participants with domain expertise will notice, and the credibility cost is higher than the cost of not knowing. Equally, avoid the vague deflection ("Great question, let's explore that later") that never actually returns to the question. If you say you'll find out, find out.

The Session That's Running Out of Time

Signal: *Forty minutes remain, sixty minutes of content remains, and the group has just entered a discussion that is genuinely productive and should not be cut.*

Response: Make the tradeoff visible and put it to the group: "We have a real conversation happening here that I don't want to rush. We also have material I planned to cover. My recommendation is that we stay with this and I send you a summary of the remaining content, but that's your call, not mine." Adults who are given a choice about their own learning time almost always choose the conversation. And they remember making the choice.

Avoid: Abandon the productive discussion to "get through the material." The productive discussion IS the material, in the sense that it is where the learning is actually occurring. A session that ends twenty minutes early and produced genuine insight is more successful than one that ran to time and covered all the slides.

Virtual and Hybrid Facilitation

Virtual and hybrid facilitation deserves specific attention because it requires preparation and in-session management that differs meaningfully from in-person delivery, and because AI tools are both more practically usable (the facilitator is at a computer) and more likely to compete with presence (the facilitator is at a computer).

Virtual facilitation's primary challenge is not technology. It is the absence of the ambient social information that in-person facilitation provides: the room's energy, the body language of participants not speaking, the side conversations that reveal what the group is really thinking, the visual confirmation that someone is confused or disengaged. Without these signals, the facilitator is operating with a significant reduction in diagnostic bandwidth. Preparation that would compensate for this, building in more explicit check-ins, designing activities that generate visible output rather than invisible thinking, creating more structured voice opportunities for participants who would otherwise disappear into camera-off silence, is precisely what AI can help design before the session.

Hybrid facilitation, where some participants are in a room and others are remote, is the most technically demanding format and the one where preparation most directly determines quality. The consistent failure mode of hybrid sessions is that remote participants become an audience to an in-room experience rather than co-participants in a shared one. Designing explicitly for remote participants' full participation, not simply accommodating their presence, requires deliberate process design that AI can help generate and stress-test.

PROMPT: Virtual and Hybrid Session Design
I need to adapt [SESSION TITLE] for virtual / hybrid delivery. Original format: [IN-PERSON, DESCRIBE ORIGINAL DESIGN] New format: [FULLY VIRTUAL / HYBRID — specify how many in-room vs. remote] Platform: [ZOOM / TEAMS / WEBEX / OTHER] Session length: [DURATION] Participant count: [TOTAL, AND IN-ROOM VS. REMOTE IF HYBRID] Please redesign this session for the new format: 1. Format-specific challenges: What are the 3 most likely facilitation challenges specific to [virtual/hybrid] delivery of THIS content with THIS audience? (Not generic virtual challenges, specific to this session and group.) 2. Activity adaptations: For each activity in the original session,

propose a version that works in [virtual/hybrid] format while preserving the pedagogical purpose. Flag any activity that cannot be meaningfully adapted and should be replaced.

3. Presence-building structures: Design 2 specific structures for this session that generate visible participation from remote participants and prevent the audience dynamic that hybrid facilitation tends to produce.

4. Technical pre-checks: List the specific technical elements I should verify in the 30 minutes before this session starts, specific to [PLATFORM] and this session's activities.

5. Recovery plans: For each likely technical failure point in this session, provide a facilitation response that keeps the group engaged while the issue is resolved.

The Facilitator and the Designer: A Necessary Partnership

The relationship between the facilitator and the designer is one of the most consequential and most under-managed relationships in the L&D ecosystem. In the best versions of this relationship, the designer and the facilitator are in genuine dialogue: the designer learns from the facilitator's experience of how the program lands in the room, and the facilitator understands the design choices well enough to make intelligent adaptations rather than inadvertent corruptions.

In the worst versions, the designer hands off a guide and moves on, and the facilitator delivers something that bears a family resemblance to what was designed but has lost the specific elements that made the design work. The scenario in Event 1 is skipped because there wasn't time. The practice activity in Event 6 is replaced with a discussion because it seemed more appropriate for this group. The assessment in Event 8 is cut because the session ran long. Each decision was locally reasonable. The cumulative effect is a program that delivers content without the instructional architecture that would make that content produce learning.

INDIE builds the designer-facilitator relationship into the Implementation phase as a structural element rather than a hopeful afterthought. The facilitator preparation described in Chapter 8 includes a design briefing conversation that is not optional. The facilitator debrief described in this chapter feeds directly into the Enhancement phase. The gap between what was designed and what was facilitated is not a quality failure to be managed away. It is information to be learned from.

> **Role Callout: Facilitator and Trainer**
>
> The most important thing a facilitator can do in an AI-augmented L&D environment is to become the design team's most reliable source of ground truth about what happens when the program meets real learners in real organizational contexts. AI can analyze survey data and synthesize feedback. It cannot sit in the room and watch what happens when the Level 2 practice activity meets a group of experienced practitioners who immediately recognize that the scenario doesn't reflect how their work actually operates. That observation, documented and communicated to the designer, is what makes the Enhancement phase produce programs that improve in ways that matter. The facilitator who makes this contribution is not performing a support role. They are performing a design role, from the most privileged observational position available.

Jerome's Preparation, Revisited

The leadership program Jerome facilitated the following spring was the most complex he had run in years: a two-day residential for thirty senior leaders, many of whom knew each other well, some of whom were in competitive relationships, and several of whom had been told by their managers that attending was not optional. The design was ambitious. The facilitation requirements were significant.

He spent more time on preparation than he had for any program in recent memory, and a meaningful portion of that preparation happened in conversation with an AI tool: generating the participant dynamics he should anticipate given the group composition, thinking through the five possible ways the Day 1 closing activity might land, building a bank of anchor questions he could use if any of the major discussions stalled, developing two complete alternative approaches to the Day 2 morning session in case the group's state the previous evening had created a different context than the design assumed.

None of the specific alternatives he had prepared were needed. What he had prepared, instead, was a quality of readiness that meant he was never surprised. When the competitive dynamics between two participants surfaced in the afternoon of Day 1, he had thought through that possibility the previous week. When the Day 2 opening fell quiet in a way that felt significant rather than empty, he had a question ready that named the silence without forcing it. When a participant disclosed something personal in the closing reflection, he acknowledged it and moved on with a grace that he attributed, privately, to having thought through that scenario specifically during preparation.

At the end of Day 2, one of the senior leaders, a person known for being difficult to impress, stopped Jerome in the corridor and said that it had been the best program he had attended in the organization. Jerome thanked him. He did not mention the preparation. He did not mention the AI tool. He did not mention the thirty scenarios he had worked through and not needed.

He didn't mention them because they were not the program. The program was the room. The preparation was what allowed him to be in it.

Chapter Summary

- Facilitation is the creation of conditions under which a group can think together and construct understanding, not the delivery of content. A facilitator's success is measured by what the group produces, not by how clearly the facilitator communicated.
- AI's role in the facilitation workflow is almost entirely pre-session: expanding the range of scenarios the facilitator has thought through, generating preparation materials that would otherwise take too long to produce, and helping the facilitator internalize program designs they did not create. The facilitation itself is irreducibly human.
- The six preparation areas, audience intelligence, session purpose clarity, process design review, hard moment anticipation, opening and closing design, and content boundary mapping, each have specific AI applications that accelerate and deepen preparation without replacing the facilitator's judgment.
- Facilitating content you didn't design is increasingly common as AI accelerates design production. The AI-assisted preparation process for this situation focuses on internalizing the design logic, identifying guide gaps, and building a mental map the facilitator owns rather than follows.

- Post-session synthesis, the work that consolidates learning for participants and captures facilitation intelligence for the design team, is chronically underinvested because it is exhausting to produce after a demanding session. AI compresses this work from two hours to forty minutes, making it likely to actually happen.
- The six hard room scenarios, the expert who knows more, the dominant voice, the resistant room, the emotional disclosure, the unanswerable question, and the time-pressured session, each have specific signal-response-avoid frameworks that AI-assisted preparation can help facilitators think through before they are in the room.
- The facilitator is the design team's most reliable source of ground truth about what happens when the program meets real learners. The facilitation debrief feeds directly into the Enhancement phase and is one of the highest-value inputs the Enhancement cycle receives.

Reflection Questions

For individual journaling or team discussion.

1. Think about your preparation for the last facilitated session you delivered. Which of the six preparation areas in this chapter did you address? Which did you skip or address inadequately? In the session, which of the gaps in preparation were visible?
2. Of the six hard room scenarios, which one do you find most personally challenging, and which response comes least naturally to you? Have you been in that situation and used the "avoid" response described in this chapter? What happened?
3. The chapter argues that the productive discussion that emerges in a session IS the material, in the sense that it is where learning is actually occurring. Do you agree? Have you abandoned a productive discussion to "get through the material?" What would it take to feel confident making the alternative choice?
4. How strong is your current relationship with the designers whose programs you facilitate? Do they know what you observe in the room? Do their designs change as a result of what you observe? If the answer is no, what is preventing that feedback loop from operating?
5. Jerome's preparation was extensive and none of his specific prepared alternatives were needed. The value was in the readiness, not the scenarios. Have you experienced this in your own facilitation practice, where preparation for something that didn't happen nonetheless changed the quality of your presence in the room? What does that experience suggest about how you should think about preparation investment?

Part Three · Chapter 13

The Performance Consultant and OD Specialist

The request came on a Tuesday, as these requests always seemed to. The head of the regional operations division had sent a one-paragraph email to the L&D director: his team's error rates had risen fourteen percent over the previous quarter, two compliance incidents had been logged, and he needed training on quality standards delivered to the division by end of quarter. Please confirm you can make this happen. The L&D director forwarded it to Nadia.

Nadia had been a performance consultant for six years. She had a master's degree in organizational development and a way of reading requests for training that her colleagues sometimes described, not always approvingly, as suspicious. She was not suspicious. She was systematic. She had learned, over a decade of this work, that requests for training almost never arrived as training problems. They arrived as performance problems, sometimes dressed in training language by someone who had decided on the solution before naming the cause. She looked at the email for a moment. Error rates up fourteen percent. Two compliance incidents. End of quarter.

She had several questions. The most important one was not "what training do you need?" It was "why did the error rates rise?" Those were different questions with potentially very different answers, and the answer to the second one would determine whether training was an appropriate response to the first at all. She replied to the L&D director: "I'd like to speak with him before we commit to anything. Can you set that up?"

The Role That Asks the Question Everyone Else Assumes Is Answered

Every other chapter in Part Three addresses roles that operate after the training decision has been made, roles that design, build, deliver, and evaluate learning solutions whose basic legitimacy has been established upstream. The performance consultant and the organizational development specialist are the roles that establish, or challenge, that legitimacy.

The performance consultant's core question is one of the most important in the L&D field and the one most frequently skipped: is training the right solution to this problem? Not "what training is needed?" but "is training needed at all, and if so, for what specifically?" The OD specialist's core question is related but wider: what organizational systems, structures, and dynamics are producing the performance gap, and what interventions, of which training may be one among several, are required to address them?

These are harder questions to ask than "can we build this training by the end of quarter?" They are harder to ask because they sometimes lead to answers that stakeholders don't want to hear, because they require access and trust that must be earned rather than assumed, and because they occasionally conclude that the L&D function is not the right owner of the solution at all. A performance consultant who concludes that the problem is a management issue is doing excellent work and will not always be thanked for it.

AI has given performance consultants and OD specialists analytical capabilities that significantly expand the speed and depth of their diagnostic work. It has not changed the questions they ask. It has not changed the organizational courage those questions require. And it has not changed the relational and systemic judgment that separates a competent diagnosis from a correct one.

The Diagnostic Foundation: Gilbert's Behavior Engineering Model

Thomas Gilbert's Behavior Engineering Model, published in 1978 in "Human Competence: Engineering Worthy Performance," is the most practically useful framework available to performance consultants for a specific reason: it is the only widely adopted framework that starts from the organizational environment rather than from the individual learner. This matters because most performance gaps are primarily caused by environmental factors, unclear expectations, missing feedback, broken tools, misaligned incentives, rather than by knowledge or skill deficits in the people doing the work.

Gilbert's model organizes potential causes of performance problems into six categories across two dimensions: environmental factors (things outside the person that affect performance) and person factors (things inside the person that affect performance). Only one of his six categories, knowledge and skills, is directly addressable through training. The other five require different interventions. A performance consultant who uses Gilbert's model systematically before recommending training is a consultant who will recommend training less often, and recommend it more accurately, than one who does not.

Gilbert's Behavior Engineering Model

Only one of six root cause categories is directly addressable through training.

Environmental Factors *(usually not training-addressable)*	**Person Factors** *(potentially training-addressable)*
Data & Information Do people know what is expected of them? Do they receive timely, specific feedback on their performance? Are consequences of poor performance visible and understood?	**Knowledge & Skills** Do people know HOW to perform? This is the only category that training directly addresses. Everything else requires a different intervention.
Resources & Tools Do people have the time, materials, equipment, and systems access they need to perform? Performance gaps caused by resource limitations are not resolved by increasing competence.	**Capacity** Do people have the physical, cognitive, or emotional capacity to perform at the required level? If not, training cannot create capacity that isn't there.
Incentives & Consequences Are the right behaviors reinforced? Are there consequences for poor performance that are proportionate and applied consistently? Motivated non-performance is a management problem.	**Motives & Preferences** Do people want to perform? Motivational gaps are real but are rarely resolved by training. They require changes to the work itself, its meaning, or its conditions.

Theoretical Grounding: Systems Thinking and the Performance System

Geary Rummler and Alan Brache's work on performance systems, extended by their concept of the "white space on the org chart" where most organizational problems actually live, provides the systems thinking framework that performance consulting requires. Rummler and Brache argue that organizational performance is produced by three levels: the organization level (strategy, structure, and deployment), the process level (how work actually flows across functions), and the individual job level (where training operates). The vast majority of performance problems that arrive as training requests are actually process or organization-level problems that training at the job level cannot resolve. A performance consultant who operates only at the job level (asking only "what do individuals need to learn?") is systematically blind to the causes of most of the problems they are asked to address.

When Training Is and Isn't the Answer

The intervention mapping table below is the most direct statement of performance consulting's core diagnostic output: given a root cause category, what does and does not fix the problem. The table is designed to be used in stakeholder conversations, not as a lecture, but as a shared diagnostic tool that allows a performance consultant and a stakeholder to reach a common understanding of what the problem actually is.

Root Cause Category	Training Fixes This	What Actually Fixes This
Knowledge/skill deficit	Yes, this is training's lane	Instruction, practice, feedback, job aids, coaching
Unclear expectations	Rarely. People may learn the task but still be unclear what's expected.	Manager clarification, role definition, performance standards documentation
No feedback on performance	No. Training cannot install the feedback loop that management hasn't provided.	Management practices, peer feedback systems, performance dashboard access
Missing or broken tools	No. Knowing how to use a working tool doesn't help with a broken one.	IT infrastructure, equipment maintenance, system access provisioning
Time and workload constraints	No. Training adds to cognitive load; it doesn't reduce workload.	Workload management, process redesign, staffing decisions, prioritization

Misaligned incentives	No. People who are rewarded for doing it wrong will continue doing it wrong regardless of training.	Compensation review, performance criteria, recognition system redesign
Selection/fit mismatch	Rarely, and not sustainably. Training cannot compensate for fundamental role-person mismatch.	Hiring criteria, role redesign, internal mobility, performance management
Culture and trust deficit	No. And training mandated on top of a broken culture often makes the trust deficit worse.	Leadership behavior, psychological safety work, OD intervention, change management

The hardest conversation in performance consulting is not the one where the data is ambiguous. It is the one where the data is clear and the conclusion is unwelcome. A performance gap caused by misaligned incentives is not a training problem. Telling a senior leader that the solution to his division's quality issues is to change the incentive structure, not to deliver training on quality standards, is a professionally exposed position to occupy. It requires credibility earned through demonstrated expertise, data presented with care, and the willingness to hold a position under pressure. AI helps with the first two. It does not help with the third.

AI in the Performance Consulting Workflow

The performance consulting workflow has five stages: data gathering, analysis, diagnosis, recommendation, and implementation support. AI has meaningful applications in all five, with the most significant impact in the first two, where the volume of data that must be synthesized before a diagnosis is possible has historically limited how thoroughly consultants could work.

Data Gathering and Synthesis

A thorough performance consulting engagement gathers data from multiple sources: performance metrics, stakeholder interviews, observation of work in practice, document review (job descriptions, process maps, standard operating procedures, prior performance reviews), and sometimes survey data from the affected population. Synthesizing this data manually, identifying the patterns across sources, finding the dissonances between what different stakeholders say, building the evidence base for a root cause hypothesis, is time-consuming work. AI compresses the synthesis without replacing the judgment about what the patterns mean.

PROMPT: Performance Data Synthesis
I am conducting a performance consulting engagement for [ROLE/DEPARTMENT] at [ORGANIZATION TYPE]. The presenting problem is: [DESCRIBE]. I have gathered the following data: Performance metrics: [PASTE OR DESCRIBE — error rates, productivity, quality scores, time-to-proficiency, etc.] Stakeholder interview themes: [SUMMARIZE KEY THEMES FROM INTERVIEWS] Observation notes: [DESCRIBE WHAT YOU OBSERVED IN THE WORK ENVIRONMENT] Document review: [DESCRIBE RELEVANT DOCUMENTS REVIEWED] Survey/pulse data: [PASTE OR DESCRIBE IF AVAILABLE] Using Gilbert's Behavior Engineering Model, analyze this data to: 1. Classify potential causes across the six BEM categories: Environmental: Data/Information | Resources/Tools | Incentives/Consequences Person: Knowledge/Skills | Capacity | Motives/Preferences 2. For each category where evidence exists, rate the strength of evidence (Strong / Moderate / Weak / Absent) 3. Identify the 2-3 most likely primary causes based on the available evidence, with the specific data points that support each 4. Identify the gaps in the current data that would most change the analysis, and recommend how to close them 5. Reach a preliminary conclusion: Is a training intervention appropriate? If yes, for which specific knowledge/skill gaps?

If no, or not primarily, what interventions are indicated?

Flag any dissonance between what different stakeholders say and what the performance data shows. Those dissonances are often where the real diagnosis lives.

Root Cause Hypothesis Testing

A diagnostic hypothesis is only as useful as the questions it generates. Once a preliminary root cause hypothesis has been formed, the performance consultant needs a set of targeted questions that will confirm or disconfirm it, questions designed to find the evidence that would prove the hypothesis wrong, not just the evidence that supports it. This is the discipline of diagnostic rigor, and it is where performance consulting separates from needs assessment: a needs assessment is designed to identify what training to build. A root cause diagnosis is designed to determine whether training is indicated at all.

PROMPT: Root Cause Hypothesis Testing

I have formed a preliminary root cause hypothesis for a performance gap in [ROLE/DEPARTMENT]: [DESCRIBE YOUR HYPOTHESIS].

Current evidence supporting this hypothesis: [LIST KEY EVIDENCE]
Current evidence that could challenge this hypothesis: [LIST IF ANY]

Please help me test this hypothesis rigorously by:

1. Generating 5-7 questions I should ask in my next stakeholder conversations that are specifically designed to DISCONFIRM the hypothesis (not to confirm it). Good hypothesis testing looks for disconfirming evidence.

2. Identifying 3 alternative hypotheses that the current evidence could equally support, and what additional data would distinguish between them

3. Generating the "if this is true, then we should see" chain for my current hypothesis: what observable evidence in the

work environment would confirm that this is the primary cause? What should be present that would be absent if a different cause were primary?

4. Identifying the stakeholder whose perspective I most need and have least accounted for in this analysis

Note: The goal is diagnostic rigor, not confirmation. I want to find the hypothesis that best explains all the evidence, not the one that is easiest to defend.

Recommendation Development and Stakeholder Communication

The performance consulting recommendation is one of the most politically sensitive documents in the L&D practitioner's repertoire. It may tell a senior leader that their team's performance problem is not a knowledge gap but a management gap. It may tell an HR director that the onboarding program they commissioned is the wrong solution to the problem they've identified. It may recommend interventions that are outside the L&D function's scope, requiring collaboration or referral to other organizational functions. Getting the recommendation right, grounded in data, clearly argued, specific about what is and is not training-addressable, and presented in a way that the stakeholder can receive, is the culmination of the entire consulting engagement.

AI can draft the analytical sections of this recommendation quickly and clearly. The judgment about what the recommendation should say, and equally importantly, how to say it to this particular stakeholder in this particular organizational context, is the performance consultant's irreplaceable contribution.

PROMPT: Performance Consulting Recommendation Report

I need to write a performance consulting recommendation for [STAKEHOLDER] regarding the performance gap in [ROLE/DEPARTMENT].

Diagnosis summary:
Primary root cause: [DESCRIBE]
Contributing factors: [DESCRIBE]

Evidence base: [SUMMARIZE KEY EVIDENCE]
Training-addressable proportion: [ESTIMATE % OF GAP THAT IS KNOWLEDGE/SKILL]
Proposed intervention portfolio:
Training recommendation (if any): [DESCRIBE SPECIFICALLY]
Non-training interventions: [DESCRIBE EACH]
Responsible parties for non-training interventions: [LIST]
Estimated timeline: [DESCRIBE]
Stakeholder context:
Stakeholder role and decision authority: [DESCRIBE]
Stakeholder's initial assumption about the cause: [DESCRIBE]
How close or far is my recommendation from their expectation: [DESCRIBE]
Organizational sensitivities I need to navigate: [DESCRIBE]
Please draft a recommendation report that:

1. Opens by naming the performance gap in measurable, business terms (not L&D terms)
2. Presents the diagnostic findings organized by root cause category with the evidence for each
3. Explains clearly which portion of the gap training can address and why the rest requires different interventions
4. Presents the full intervention portfolio with owner, timeline, and expected impact for each element
5. Closes with a specific request: what decision or action does this stakeholder need to take?

Tone: Evidence-based and direct, without being confrontational. Assume an intelligent stakeholder who wants to solve the problem and may not have considered that training isn't the whole answer.

Role Callout: Performance Consultant

The most important professional development investment for a performance consultant in an AI-augmented environment is not analytical skill; AI has dramatically accelerated the analytical work. It is the ability to have the difficult conversation: to tell a senior leader that the solution to their performance problem is not training, and to say so in a way that builds rather than damages the relationship. This requires credibility established over time, an evidence base that makes the diagnosis hard to dismiss, and the professional confidence to hold a position under pressure. AI can help build the evidence base. The credibility and the confidence are earned through practice and cannot be accelerated by any tool.

The OD Specialist in INDIE

The organizational development specialist operates at a broader level than the performance consultant, not just diagnosing why specific performance gaps exist, but examining the organizational systems, structures, and culture that produce or prevent effective performance across the organization. Where the performance consultant asks, "why isn't this working?" the OD specialist asks, "what kind of organization do we need to be, and what does it take to become it?"

The OD specialist's work is more organizational than it is instructional, and the INDIE framework applies to it somewhat differently than to the roles covered in previous chapters. The OD specialist uses the IN phase's analytical tools for organizational diagnosis rather than individual needs analysis. They may use the Development phase's content creation capabilities to produce change communications, leadership development materials, or culture-building resources rather than training courses. They use the Implementation phase's deployment thinking for change rollout rather than program launch. And they use the Enhancement phase's measurement infrastructure to track organizational change rather than training effectiveness.

What is consistent across both roles, performance consulting and OD, is that their most important work is diagnostic and systemic, and that their most important contribution is the analysis that determines what interventions are needed before anyone begins building them. This is where both roles are most powerful and most exposed: the diagnosis that recommends against training, or that names a leadership problem, or that concludes the organization's structure is producing the problems everyone is trying to train their way out of.

OD Focus Area	AI Accelerates	Human Judgment Remains Essential
Change Readiness Assessment	Synthesizing employee sentiment data, pulse survey analysis, stakeholder interview themes, and change history patterns into a readiness profile	Reading the informal power structure, assessing whether stated readiness reflects genuine willingness or compliant lip service, and deciding what the data means for timing

Culture Diagnosis	Analyzing engagement survey data, exit interview patterns, and qualitative culture research at scale; generating hypotheses about the gap between espoused and enacted values	Distinguishing symptoms from causes in culture data; making the judgment call about whether the culture problem is a leadership problem, a systems problem, or both; navigating the political sensitivity of the findings
Organizational Design	Mapping current organizational structure, identifying spans of control and coordination mechanism patterns, generating design alternatives and their tradeoffs	Understanding the informal organization that the org chart doesn't show; assessing which design alternatives are politically feasible; reading what the organization's history means for future design choices
Leadership Development	Generating development pathways, 360 feedback synthesis and pattern analysis, coaching question banks, and leadership competency content	Assessing genuine developmental readiness in individual leaders; making the judgment about which development interventions match which developmental challenges; managing the coaching relationship itself
Team Effectiveness	Analyzing team interaction data, generating team charter templates, producing conflict resolution frameworks and process improvement tools	Reading team dynamics in real time, understanding what is named and what is not named in a team's conversation, facilitating the difficult conversations that team effectiveness often requires
Strategic Planning Support	Environmental scanning synthesis, scenario planning frameworks, competitive analysis structuring, strategic assumption documentation	Integrating quantitative analysis with qualitative organizational knowledge; assessing strategic feasibility given the organization's actual capabilities and culture; facilitating the decision-making process among senior leaders with competing agendas

OD and Change Management: AI in the Change Process

Change management, the structured approach to transitioning individuals, teams, and organizations from a current state to a desired future state, is one of the OD specialist's core competencies and one of the areas where AI has the most significant application. The data-intensive work of change management: assessing readiness, segmenting stakeholder populations, designing communication strategies, monitoring adoption, is precisely the kind of analytical and generative work that AI handles well.

What AI cannot do in change management is understand the informal organization, the web of relationships, loyalties, grievances, and informal power that determines whether a change initiative succeeds or becomes another organizational memory of failed transformation. The change plan that makes perfect sense on paper fails because the three people who actually shape opinion in the affected team are not on anyone's stakeholder map. The communication strategy that receives enthusiastic formal approval is quietly undermined by a manager who disagrees with the direction and has the informal credibility to make his disagreement contagious. These dynamics are invisible to AI and visible only to people who have spent time in the organization and have earned enough trust to see what is not on the org chart.

PROMPT: Change Readiness Assessment
I am assessing change readiness for [DESCRIBE THE CHANGE INITIATIVE] in [DESCRIBE THE ORGANIZATION / AFFECTED POPULATION]. Available data: Pulse/engagement survey data: [PASTE OR DESCRIBE] Stakeholder interview themes: [SUMMARIZE] Prior change history in this organization: [DESCRIBE RELEVANT HISTORY] Known resistance sources: [DESCRIBE IF IDENTIFIED] Timeline and pace of the change: [DESCRIBE] Please produce a change readiness analysis that: 1. Assesses readiness across three dimensions: - Awareness: Do people understand why this change is happening? - Desire: Do people want to support and participate in it?

- Ability: Do people have what they need to change?
Rate each dimension (High / Moderate / Low) with evidence.

2. Segments the stakeholder population by readiness profile:
   - Champions (high awareness, high desire, high ability)
   - Fence-sitters (moderate or unformed position)
   - Resisters (low desire regardless of awareness or ability)
   - Silent majority (passive, not yet engaged)

3. Identifies the 3 highest-risk readiness gaps and proposes a specific action to address each

4. Flags what the available data does NOT tell me — what I would need to learn from direct organizational observation that no data source can provide

Note: The fourth item is often the most important. The informal organization is not visible in survey data.

PROMPT: Making the Case for OD Investment

I need to make a business case for an organizational development intervention to [DESCRIBE STAKEHOLDER AUDIENCE].

The organizational problem: [DESCRIBE THE PERFORMANCE OR CULTURE PROBLEM]
The diagnosis: [SUMMARIZE THE ROOT CAUSE ANALYSIS]
The proposed OD intervention: [DESCRIBE]
Why training alone is insufficient: [DESCRIBE THE SYSTEMIC FACTORS]
Available evidence: [DESCRIBE PERFORMANCE DATA, COST DATA, BENCHMARKS]
Estimated investment required: [DESCRIBE]

Please build a business case that:

1. Opens by naming the cost of the current state in terms the stakeholder audience cares about (revenue, retention, risk, customer impact — not L&D metrics)
2. Explains clearly why this is a systems problem, not a training problem, in language that doesn't require OD expertise to understand
3. Describes the proposed intervention and how it addresses the systemic causes
4. Presents the investment required alongside a credible estimate

of the cost of inaction
5. Closes with a specific decision request

Avoid: L&D and OD jargon. Assume a financially literate audience with no background in organizational development who needs to understand why this is a leadership and systems problem, not a training gap that a well-designed course would close.

The Courage Competency

The AI tools available to performance consultants and OD specialists in 2026 are genuinely powerful. They can synthesize data at a scale that was previously impractical, generate diagnostic frameworks that organize complex organizational information, produce intervention recommendations that are structurally sound, and draft the reports and presentations that communicate findings to senior audiences. None of this changes the most important competency both roles require: the willingness to say what the data actually shows, even when it is unwelcome.

Performance consulting and OD practice have a specific temptation that AI makes worse rather than better: the temptation to produce a recommendation that the client will accept rather than a recommendation that the data supports. AI tools are trained on human-generated data, which includes a great deal of organizational communication designed to be agreeable rather than accurate. An AI-generated recommendation report asked to be "diplomatic" will produce something diplomatic. It may not produce something true.

The discipline of honest diagnosis is a human discipline. It requires the practitioner to hold the data above the relationship when the data and the relationship are in tension, to say "the evidence points to a management problem" when the person in the room is the manager, to say "training is not the right solution here" when the person in the room is the person who requested training and has already told their team it's coming, to say "this change initiative is at high risk of failure" when the person in the room has staked their credibility on its success.

The practitioner who does this well is not being brave for its own sake. They are serving their client's actual interests rather than their stated preferences. The most useful thing a performance consultant can do for a senior leader who is about to spend significant organizational capital on a training solution to a management problem is to say so, clearly, while there is still time to do something else. That is the service. The courage to provide it is the competency.

> **Role Callout: Performance Consultant and OD Specialist**
>
> AI gives you the ability to do more analytical work faster than was previously possible. Use that capacity to build stronger diagnostic cases, not to produce more recommendations faster. A performance consulting engagement that concludes in three weeks with a strong, data-grounded diagnosis is more valuable than one that concludes in one week with a plausible-sounding recommendation that hasn't been tested rigorously. AI accelerates the production of recommendations. It does not improve the rigor of the thinking behind them. That rigor is your professional contribution, and it cannot be compressed without quality cost.

Nadia's Conversation

The call with the head of regional operations lasted forty minutes. Nadia had spent an hour before running the available data through an AI synthesis tool: the error rate trends by team and by manager, the compliance incident reports, the most recent employee engagement pulse for the division, and three months of performance review summaries that the L&D director had pulled from the HRIS (Human Resources Information System). The AI synthesis had produced a preliminary hypothesis before she picked up the phone.

The hypothesis was: the performance gap was not uniformly distributed across the division. It was concentrated in two of the seven regional teams. The other five teams' error rates had held steady. The two affected teams had both changed managers in the previous six months. Both new managers had come from technical roles without prior people management experience. The engagement pulse for those two teams was the lowest in the division. She had tested the hypothesis with three disconfirming questions before the call. None of them had found evidence against it.

In the call, she asked the divisional head to describe the distribution of the errors. He paused. "Actually, now that you ask, most of it is coming from the northeast and the coastal teams." She asked about those teams' recent history. Another pause. "We put two new managers in. They're strong technically. Why?"

She told him what the data showed. She told him what it suggested about the cause. She told him that quality standards training delivered to the whole division would address, at best, a small portion of the problem in two of the seven teams, and would cost the division time and attention that could be better spent on new manager coaching and structured onboarding support for the affected teams' supervisors. There was a longer pause. "That's not what I asked for." "No," Nadia said. "But I think it's what you need." There was a silence that she had learned, over eleven years, to let sit. Then: "What would the manager coaching look like?"

She had a draft answer ready. The AI tool had helped her produce three possible intervention designs before the call. She walked him through the one that fit the timeline and the organizational context. By the end of the call, they had a plan that did not include a division-wide training program on quality standards. It included something that would actually work

Chapter Summary

- The performance consultant and OD specialist are the roles that ask whether training is the right solution, the question that every other L&D role assumes has been answered. Their most important contribution is the diagnosis that precedes all design, not the design itself.
- Gilbert's Behavior Engineering Model organizes performance causes into six categories across two dimensions. Only one category, knowledge and skills, is directly addressable through training. Systematic application of the model before recommending training produces fewer training recommendations and more accurate ones.
- The intervention mapping table makes the performance consultant's core argument concrete: most performance gaps have multiple causes, only some of which are training-addressable. Telling a stakeholder this clearly, with evidence, is the service. The courage to provide it is the competency.
- AI's most significant applications for performance consultants are in data synthesis (collating and pattern-finding across multiple data sources), hypothesis testing (generating disconfirming questions and alternative explanations), and recommendation drafting. AI does not improve the rigor of the diagnostic thinking behind the recommendations.
- The OD specialist applies the same diagnostic orientation to organizational systems, structures, and culture. AI accelerates OD's data-intensive work

(readiness assessment, culture diagnosis, change communication) while the invisible organization, informal power, unspoken dynamics, what is not on the org chart, remains exclusively visible to people who have earned the trust and time to see it.

- The courage competency is the one AI cannot supply and cannot accelerate: the willingness to say what the data shows when it is unwelcome, to hold the diagnosis above the relationship when they are in tension, and to serve the client's actual interests rather than their stated preferences.

Reflection Questions

For individual journaling or team discussion.

1. Apply Gilbert's Behavior Engineering Model (BEM) to the last three training requests you received or acted on. For each, in which BEM categories did the evidence suggest the primary cause lived? Was the intervention delivered actually matched to those causes, or was training delivered for a problem in a different category?
2. Think of a time you recommended training when, in retrospect, the evidence pointed to a non-training cause. What made it easier in that moment to proceed with training than to name the actual cause? What would it have taken to have Nadia's conversation instead?
3. The chapter argues that AI makes the temptation to produce agreeable recommendations worse rather than better. Have you experienced this, where an AI tool produced a recommendation that was diplomatic rather than accurate? What would a "rigor over diplomacy" prompt look like for the recommendations you produce?
4. Looking at your current organizational context: which performance gaps that are being addressed through training are actually being caused by factors in Gilbert's environmental categories, unclear expectations, missing feedback, broken tools, misaligned incentives? What prevents those systemic causes from being named?
5. The OD row in the chapter's table notes that the informal organization is invisible to AI and visible only to those who have earned enough trust to see it. What is the informal organization in your current work context, the power, relationships, and dynamics not on any org chart, that is most relevant to the performance and change work you are trying to do? How did you come to see it?

Part Three · Chapter 14

The L&D Manager and Curriculum Coordinator

The quarterly L&D review was in forty minutes. Claudette had been the L&D director for three years, and she had not yet found a version of this meeting she completely enjoyed. The CHRO (Chief Human Resources Officer) wanted evidence of impact. The business unit heads wanted programs that addressed their specific problems, delivered on their specific timelines, by a team that was, in her CHRO's view, already slightly too large for what it was producing. Two members of her team were experimenting with AI tools independently and producing work that was faster but variable in quality. One team member was actively resistant and had told her privately that she thought AI was going to make training worse, not better. The team's skill levels were uneven in ways that had always been manageable and were now, with AI changing the production landscape, becoming a more visible gap.

On her desk was a spreadsheet she had made the previous evening: forty-three programs in the portfolio, with the last time each had been reviewed, the last Enhancement data collected, and a rough estimate of the organizational priority each served. Of the forty-three, she had confidence in twelve. Another fifteen were solid but needed work. Sixteen were programs she had inherited or agreed to build under pressure and had never had the bandwidth to examine critically. She didn't know whether they were working.

The CHRO's opening question at these reviews was always some version of the same thing: "What are we getting for what we're spending?" She had never had a fully satisfying answer. She had good answers for the twelve programs she was confident in. She had incomplete answers for the fifteen. For the sixteen, she had almost nothing. She closed the spreadsheet, picked up her notes, and thought, "this is the year I fix that."

The Manager's Vantage Point

The L&D manager's job is structurally different from every other role in this section of the book. The practitioner roles covered in Chapters 10 through 13 each focus on a specific type of work: designing, building, facilitating, consulting. The manager's job is to make all of that work possible, to ensure that the right work is happening, that the team doing it has the capability and the resources to do it well, that the organizational relationships that give L&D its mandate are maintained and developed, and that the portfolio of programs the team produces is coherent, effective, and aligned with what the organization actually needs.

This is a different relationship to quality than the practitioner roles have. An instructional designer can be responsible for the quality of the programs they design. An L&D manager is responsible for the conditions that make quality possible across a team and a portfolio, which is a harder problem, because it requires influencing and enabling rather than doing, and because the quality failures that result from poor management conditions are often invisible until they have been accumulating for months.

The curriculum coordinator occupies a related but distinct position: responsible for the coherence and currency of a curriculum rather than the management of a team. In some organizations this is a junior role supporting a broader L&D function; in others, particularly in higher education, healthcare, and large professional certification bodies, it is a senior strategic role with significant authority over what the organization teaches and how. This chapter addresses both roles because their core challenges are analogous: managing at the portfolio level, making decisions about programs as a collection rather than as individual artifacts, and maintaining quality standards across a body of work that is always larger than any one person can hold in their attention simultaneously.

Portfolio Management in an AI-Augmented Environment

The forty-three programs in Claudette's portfolio represent a problem that is common to L&D teams of any meaningful size: the portfolio grows faster than the capacity to evaluate it. Programs are built in response to specific organizational needs, delivered, and then maintained by inertia, updated occasionally when someone remembers or a compliance requirement forces it, evaluated intermittently when time allows, and retired rarely because retirement requires a decision that nobody wants to own.

The result is a portfolio that represents an organization's historical learning needs more than its current ones, that has uneven quality across programs, and that consumes maintenance bandwidth that could be invested in better design and genuine enhancement. AI does not fix this problem. What AI can do is make the problem visible faster and more completely, surfacing the programs that deserve scrutiny, providing the analytical foundation for portfolio decisions, and compressing the audit work that would otherwise take weeks.

Portfolio Health Matrix

Five dimensions for evaluating program and portfolio health, with AI applications and manager judgment calls.

Portfolio Dimension	Questions to Ask	AI Can Surface	Signals of Health	Signals of Risk	Manager's Call
Strategic alignment	Does each program connect to a current organizational priority? When did we last review alignment?	Patterns in which programs receive strong engagement vs. low completion; programs with no recent stakeholder champion	Programs regularly cited in business reviews; completion rates correlate with business metric movement	Programs requested years ago for problems that have since changed; low completion with no stakeholder concern	Retire or redesign programs that no longer serve current priorities, even if they were expensive to build

Coverage and gaps	What competencies does our portfolio address? What doesn't it touch that the organization needs?	Competency map comparison against portfolio; frequency analysis of skills noted in performance reviews not addressed by current programs	Portfolio covers the competency framework without significant gaps; emerging skill needs are identified before they become critical	Reactive program building in response to crises; same competency addressed by multiple overlapping programs	Prioritize gap-filling based on business impact, not urgency; resist building duplicative solutions
Quality and effectiveness	Which programs produce demonstrated learning and transfer? Which ones have never been evaluated beyond Level 1?	Level 1–4 data aggregated across portfolio; programs with persistent low scores or no transfer data	Strong Level 2 and 3 data across core programs; enhancement cycles actively running on high-priority content	Programs with only Level 1 data after multiple cycles; programs with known transfer problems that have never been addressed	Allocate enhancement resources to high-impact programs; sunset programs that cannot demonstrate value after fair investment
Currency	When was this content last reviewed against current practice? Have the job requirements it addresses changed?	Programs not updated in 24+ months flagged for review; AI scan of content against current published standards	Regular content review cycle operating; programs updated proactively when context changes	Programs with outdated examples, superseded regulations, or technology references to tools no longer in use	Distinguish content that requires full redesign from content that requires targeted update; avoid treating all aging

		or practices			content the same
Team capacity fit	Does our current team have the capability to deliver, maintain, and enhance this portfolio?	Portfolio load analysis: hours required vs. team capacity; skills inventory gap against portfolio requirements	Portfolio scale matches team capacity with room for quality investment; team skills cover all portfolio needs	Team stretched thin maintaining existing content with no capacity for enhancement or new development	Make the capacity argument to leadership with data; portfolio scope must match team resources or quality will degrade invisibly

PROMPT: Portfolio Health Audit
I need to conduct a health audit of my L&D program portfolio. Portfolio inventory: [PASTE OR DESCRIBE — program names, dates last reviewed, available evaluation data, organizational priority served] Organizational context: Current strategic priorities: [LIST 3-5] Team capacity (FTE available for maintenance and development): [DESCRIBE] Most recent CHRO or leadership feedback on L&D priorities: [DESCRIBE] Please conduct a portfolio health analysis across five dimensions: 1. Strategic alignment: Which programs clearly serve current priorities? Which are misaligned or no longer connected to active business needs? 2. Coverage and gaps: What competency areas are over-served (multiple programs) and under-served (real needs with no current program)? 3. Quality and effectiveness: Which programs have strong evaluation data? Which have no data beyond Level 1, or persistent quality concerns? 4. Currency: Which programs have not been reviewed in 24+ months, or address requirements that have since changed? 5. Capacity fit: Given team capacity, which programs are consuming disproportionate maintenance bandwidth relative to their impact?

Produce a prioritized list of:
- Programs to invest in (strong alignment, capacity available)
- Programs to maintain (solid but not high-priority for enhancement)
- Programs to review (uncertain value; need evaluation before decision)
- Programs to retire (clear misalignment, low value, or no evidence)

Flag any program where the retirement decision may have political dimensions I should anticipate and plan for.

PROMPT: Making the Case for Portfolio Rationalization

I need to make a case to [CHRO / HR LEADERSHIP] for rationalizing our L&D portfolio — specifically for retiring or consolidating [NUMBER] programs that are no longer serving current priorities.

Programs under consideration for retirement/consolidation: [LIST WITH brief description of each and rationale for the decision]
Business case for the change: [DESCRIBE — freed capacity, quality investment, strategic focus]
Likely objections: [DESCRIBE WHO WILL PUSH BACK AND WHY]
Current utilization data: [COMPLETION RATES, LAST DELIVERY DATE, ETC.]

Please draft a portfolio rationalization proposal that:
1. Opens by naming the opportunity cost of the current state (capacity consumed maintaining programs of uncertain value)
2. Presents the retirement/consolidation recommendations with the evidence for each decision
3. Connects freed capacity to specific higher-value investments
4. Anticipates and addresses the likely objections
5. Proposes a transition approach that respects stakeholder relationships (notifying program owners, archiving materials, etc.)

Tone: Business-oriented, evidence-grounded, respectful of the work that went into the programs being retired. These are decisions about resource allocation, not verdicts on past quality.

Team Capability Development in an AI-Augmented World

The L&D manager's responsibility for team capability has always included keeping practitioners current with evolving tools and methods. What is different now is the pace of change, the unevenness of adoption across team members, and the specific quality risks that AI-augmented workflows introduce. A team where some members are using AI tools with rigor and others are not, or where some are using them and others are actively avoiding them, is a team producing work of systematically different quality without a shared standard for what that difference looks like.

The manager's task is not to mandate uniformity of tool adoption. It is to establish shared standards for quality that apply regardless of which tools are used, and to build the team's collective capacity to use AI tools in ways that meet those standards. This is a development challenge, not a compliance challenge, and it requires the same deliberate design that the team applies to any other learning initiative.

Assessing Where the Team Is

The AI maturity assessment below maps five levels of team AI practice from unaware to leading. The manager's first task is an honest assessment of where the team currently sits, which is rarely uniform. In most teams, there is a distribution: a few practitioners at Level 2 or 3, most at Level 1, and perhaps one or two who are actively resistant or genuinely at Level 0. Managing this distribution requires different approaches for different positions on the scale.

Team AI Practice Maturity Assessment

Five levels of team AI practice, with the manager action required at each.

Level 0: Unaware	Level 1: Experimenting	Level 2: Adopting	Level 3: Integrating	Level 4: Leading
AI tools unknown or actively avoided. No shared language about AI in L&D practice. Team skepticism is high or unaddressed.	Individuals using AI tools independently, often without team awareness. No shared standards. Quality variable and unchecked.	AI use is explicit and acknowledged. Beginning to develop shared prompts and quality standards. Some team members more advanced than others.	AI integrated into standard workflows. Shared prompt library in use. Quality review of AI output is routine. Team capability broadly similar.	Team actively develops and shares AI practices. Contributes to organizational AI governance. Continuously improves workflows. Acts as internal resource for other functions.
Manager action: Create safety and shared language. Start with one honest team conversation about what people actually know and fear.	Manager action: Surface and share what's happening. Acknowledge the experimentation positively. Begin building shared standards from what's working.	Manager action: Formalize prompt library. Build peer learning into team rhythm. Close the gap between early adopters and those still behind.	Manager action: Invest in advancing the leading practitioners. Begin contributing to organizational AI governance conversations.	Manager action: Sustain the culture. Protect the team's learning time. Avoid letting operational pressure erode the investment in continuous improvement.

Building the Shared Prompt Library

The single most practical team capability investment an L&D manager can make is building and maintaining a shared prompt library, a documented collection of AI prompts that the team has tested, refined, and collectively endorsed as producing quality-meeting output for specific tasks. A shared prompt library does three things at once: it accelerates onboarding for new team members, it ensures quality consistency across practitioners at different skill levels, and it creates a living record of the team's collective AI learning that compounds over time.

The library is not a static document. It requires governance: a process for adding new prompts (tested and reviewed, not simply contributed), a process for retiring prompts that no longer produce adequate results as model versions change, and a clear ownership structure so the library is maintained rather than allowed to grow into an unmanageable archive. The Appendix of this book contains a starter prompt library organized by INDIE phase and task. It is a beginning, not a ceiling.

PROMPT: Building a Team Prompt Library Framework
I need to establish a shared AI prompt library for my L&D team of [NUMBER] practitioners with the following mix of roles: [LIST ROLES]. Current state: AI tool usage on the team: [DESCRIBE which tools, how widely used] Quality consistency: [DESCRIBE CURRENT VARIATION IN AI-ASSISTED WORK] Existing documented prompts: [DESCRIBE IF ANY] Please design a prompt library framework that includes: 1. Organization structure: How should prompts be categorized? (By INDIE phase, by role, by task type, by output format) Recommend the structure that will be most navigable for this team. 2. Contribution standards: What must be true about a prompt before it enters the library? - Testing requirement (how many times used, by whom) - Documentation requirement (prompt text, intended use, known limitations, quality review steps required) - Approval process 3. Maintenance cadence: How often should prompts be reviewed? What triggers a review (model updates, quality complaints, workflow changes)? 4. Governance: Who owns the library? Who can add, modify, retire? 5. Onboarding integration: How does a new team member learn to use the library? What is the minimum they should know before using library prompts in production work? Also suggest the first 5 prompts this team should prioritize documenting based on the role mix described.

Managing the Resistor and the Zealot

Every team navigating an AI transition contains two people who represent opposite failure modes. The resistor, the practitioner who is actively avoiding AI tools, perhaps publicly skeptical, perhaps quietly non-compliant with team expectations, is the more visible challenge and often the one managers spend the most time on. The zealot, the practitioner who is enthusiastically adopting every new AI tool, producing high volumes of AI-assisted work, and sometimes bypassing quality review in the name of speed, is the more dangerous one, because the quality problems they produce are less visible until they reach a stakeholder.

Managing both requires the same fundamental approach: a clear, shared quality standard that neither resistance nor enthusiasm can override. For the resistor, the conversation is: "I respect your concerns, and some of them are legitimate. The team's quality standards apply to all work, regardless of tools, and those standards include the expectation that you can use the tools the team has adopted for the tasks they serve well." For the zealot, it is: "Your energy is valuable. The quality review steps exist because AI output requires expert evaluation before it meets our standards, and skipping them is not a speed optimization; it is a quality debt that we will pay later."

Both conversations are easier to have when the quality standard has been established explicitly and collectively, when the team has agreed, in advance, what quality means for AI-assisted work. A standard the team built together is harder to resist or circumvent than one that was imposed from above, and it gives the manager a shared reference point rather than a personal opinion.

> **Role Callout: L&D Manager**
>
> The most important structural change an L&D manager can make in an AI-augmented environment is to make quality review of AI output a named, non-negotiable workflow step rather than a personal responsibility left to each practitioner's judgment. This means defining what quality review consists of (the criteria, not just the instruction), building it into project timelines as a milestone, and treating AI output that has not been reviewed the same way you would treat any other deliverable that bypassed the quality gate: not as fast work but as incomplete work.

Introducing INDIE to a Team That Has Used ADDIE for Years

For many L&D managers reading this chapter, the practical challenge is not understanding INDIE; it is introducing it to a team that has built its professional identity around ADDIE and its stakeholder relationships around ADDIE's deliverable structure. This is a change management problem inside an L&D team, and it benefits from the same systematic approach that any organizational change requires.

The 90-day roadmap below is a structured approach to INDIE adoption that has been designed for teams at various levels of current AI maturity. It is not a rigid plan but a scaffold: the activities at each phase are the minimum viable steps for building genuine adoption rather than nominal compliance. A team that completes the activities in name only, that has an INDIE rollout meeting and then continues working exactly as before, has not adopted INDIE. It has performed adoption.

90-Day INDIE Adoption Roadmap

A structured approach for L&D managers introducing INDIE to teams currently using ADDIE.

Phase 1 *Weeks 1–3*	**Foundation: Shared Understanding Before Shared Practice** Do not begin with tools or prompts. Begin with the question: what problems does our current process create that INDIE addresses? Run one team session (ninety minutes) in which each team member describes a recent project where ADDIE's linear structure created friction. Use those examples to build a shared, team-specific case for why the framework needs to evolve. Assign Chapter 5 of this book as pre-reading. The goal of Phase 1 is not commitment to INDIE; it is honest acknowledgment of ADDIE's real limitations in this team's actual work.
Phase 2 *Weeks 4–6*	**Pilot: One Project, Full INDIE, Real Conditions** Select one active project, medium-sized, real stakes, willing stakeholder, and run it explicitly through the INDIE framework. Assign a design lead who is enthusiastic about the transition. Run weekly thirty-minute reflection sessions with the project team: what is working, what is harder than expected? Document the process, not just the output. The pilot is not a showcase of INDIE working perfectly. It is a learning event for the team, and the most valuable outcome is a specific, honest account of where the framework created value and where it required adjustment.

Phase 3 *Weeks 7–9*	**Standard: Shared Language and First Library Entries** After the pilot, run a team retrospective: what did we learn that should change how we work? Codify the answers into a team INDIE standard, not a document that copies the framework verbatim, but a team-specific account of how INDIE applies to this team's context, clients, and constraints. Begin the shared prompt library with the prompts the pilot team found most useful and most reliable. Establish the quality review standard for AI-assisted work. These three artifacts, team standard, prompt library, quality review protocol, are the infrastructure on which adoption runs.
Phase 4 *Weeks 10–12*	**Scale: All Projects, Sustainable Rhythm** Expand INDIE application to all active projects. Hold a monthly thirty-minute team learning session where practitioners share one thing that worked, one thing that didn't, and one prompt or technique they want to share with the team. This session is not a status update. It is a community of practice in miniature, and it is what sustains adoption past the initial enthusiasm. At the end of Week 12, run a team assessment against the AI maturity framework. Where did the team start? Where is it now? What is the next development priority?

PROMPT: INDIE Adoption Stakeholder Communication

I need to communicate to my internal stakeholders that my L&D team
is transitioning from ADDIE to an updated design framework called INDIE.

Audience: [DESCRIBE CHRO, business unit heads, HR business partners, etc.]
Current state: Team has used ADDIE for [NUMBER] years. Stakeholders are familiar with ADDIE's deliverable structure (analysis documents, storyboard reviews, implementation plans, evaluation reports).
What changes for stakeholders: [DESCRIBE SPECIFICALLY timeline, deliverable structure, review points, etc.]
What doesn't change: [DESCRIBE quality standards, stakeholder involvement, evaluation rigor, etc.]

Please draft a stakeholder communication that:

1. Opens with why the change is happening, framed as organizational benefit (faster delivery, better alignment, continuous improvement) not as team process improvement
2. Explains what will be different in how the team works with them
3. Explicitly names what will not change (the things stakeholders most worry about losing)
4. Describes what good looks like under the new framework (so they know what to expect and how to evaluate it)
5. Invites a conversation rather than announcing a decision.

Curriculum Coordination at Scale

The curriculum coordinator's challenge is related to but distinct from the L&D manager's: where the manager's portfolio problem is about organizational alignment and team capacity, the coordinator's curriculum problem is about coherence, sequencing, and currency across a body of learning content that may span dozens of courses, multiple delivery formats, several subject matter domains, and learner journeys that unfold over months or years.

A curriculum that lacks coherence produces learners who experience contradictions between courses that claim to be part of the same program, who encounter the same content in multiple places and miss critical content in none, and who cannot see the thread that connects what they are learning to where they are trying to go. A curriculum that lacks currency produces learners who are trained on outdated content, obsolete procedures, and superseded standards, and who may not know it until the gap appears in their performance.

AI has meaningful applications for curriculum coordination work that have been largely underexploited because the coordinators who could use them most are often the practitioners with the least time to experiment. Content gap mapping, alignment verification, currency auditing, and prerequisite structure analysis are all tasks that AI can accelerate significantly, compressing work that would otherwise consume weeks into something manageable in days.

PROMPT: Curriculum Alignment and Coherence Audit
I need to audit the coherence and alignment of a curriculum. Curriculum: [NAME / DESCRIBE THE PROGRAM OR LEARNING PATHWAY] Scope: [NUMBER] courses/modules covering [DESCRIBE TOPIC AREAS] Target learner journey: [DESCRIBE WHO TAKES THIS, IN WHAT SEQUENCE, TOWARD WHAT OUTCOME] Available documentation: [PASTE COURSE DESCRIPTIONS, OBJECTIVES, OR CONTENT OUTLINES] Please conduct a curriculum coherence audit that identifies: 1. Objective alignment: Are the objectives across courses building

toward the stated curriculum outcome? Are there any courses whose objectives are disconnected from the overall pathway?

2. Sequencing logic: Does the sequence of courses reflect a coherent learning progression? Are prerequisites respected? Are there places where content is presented before the foundational knowledge to understand it has been established?

3. Coverage analysis: What competency areas are addressed by multiple courses (potential redundancy or productive reinforcement)? What competency areas appear in the objectives but are not covered by any course (gaps)?

4. Contradiction check: Are there any places where courses present conflicting approaches, terminology, or standards that a learner moving through the curriculum would experience as contradictory?

5. Currency risk: Based on the topic areas described, which courses are most likely to contain outdated content given the pace of change in this domain? Flag for priority review.

Produce a prioritized list of curriculum improvements with a suggested implementation sequence.

The Organizational Relationship: Making the Case for L&D

No amount of portfolio quality, team capability, or INDIE adoption matters if the L&D function does not have the organizational standing to do meaningful work. The L&D manager's most important non-tactical responsibility is maintaining and building the organizational relationships that give the function its mandate, the credibility with the CHRO that allows honest conversations about what the function needs, the relationships with business unit leaders that generate the strategic requests rather than just the reactive ones, and the demonstrated impact that makes the function's budget defensible.

AI changes this in a specific way: it gives L&D managers access to better data, faster, than was previously possible. The quarterly review that Claudette dreaded, the CHRO's question about what the organization is getting for what it's spending, has a better answer available when the Enhancement phase is running properly and the portfolio health matrix is current. The answer is not "here are our completion rates." It is "here are the twelve programs we have strong evidence for, here is what that evidence shows, here is what we're doing about the programs we don't have strong evidence for, and here is what we're retiring because it's not serving current priorities."

That conversation requires data. It also requires the confidence to have it honestly, to acknowledge the sixteen programs without strong evidence rather than presenting only the twelve that look good. Leadership credibility for L&D comes not from always having the right answer but from being the function that engages with evidence honestly, makes decisions based on what the data shows rather than what is comfortable, and continuously improves in ways that are visible and measurable. That is what the INDIE framework, properly implemented, enables.

PROMPT: L&D Quarterly Leadership Report
I need to prepare a quarterly L&D report for [CHRO / EXECUTIVE LEADERSHIP]. Portfolio overview: [NUMBER] active programs, [NUMBER] in development Team size: [NUMBER] FTE Investment (approximate, if shareable): [DESCRIBE] Key data available: [LIST DATA SOURCES — completion rates, Level 1-4 data, business metric connections, utilization stats] Most significant accomplishments this quarter: [LIST 3-4] Most significant challenges this quarter: [LIST 2-3] Portfolio decisions made (programs retired, redesigned, added): [DESCRIBE] Strategic priorities the function served this quarter: [LIST] Please draft a quarterly report that: 1. Opens with a one-paragraph executive summary: what did L&D accomplish this quarter that mattered to the organization, in business terms 2. Presents portfolio data organized by strategic priority served (not by program name) 3. Shows the strongest evidence of impact at Level 3 or 4 for the programs that have it

4. Honestly names the programs without strong evidence and describes the plan to address them
5. Identifies one area where the function needs investment or support to do better work, with a specific request
6. Closes with the priority focus for next quarter

Audience: Financially literate executives who value directness, evidence, and honest acknowledgment of gaps over polished summaries that avoid difficult information.

Role Callout: L&D Manager and Curriculum Coordinator

The most important shift in how L&D managers present their work to leadership in an AI-augmented environment is from reporting activity to demonstrating value. "We delivered 47 programs to 2,300 participants" is activity. "These six programs produced measurable behavior change, here is the evidence, and we retired four programs that didn't" is value. AI gives you the analytical capacity to produce the second kind of account. The willingness to present it honestly, including the programs without evidence, the retirement decisions, the resource constraints that are limiting quality, is what builds the kind of organizational trust that makes L&D a strategic partner rather than a service provider.

Claudette's Review

The quarterly review lasted fifty-five minutes. Claudette brought two things she had not brought to previous versions of this meeting: a portfolio health summary she had built using AI to synthesize three years of scattered evaluation data into a coherent view, and a retirement list.

The health summary showed twelve programs with strong evidence, fifteen with solid delivery records and incomplete evaluation data, and sixteen with neither. She did not lead with the twelve. She led with the full picture. The CHRO's first question was the expected one: "What are we getting for what we're spending?"

Claudette walked her through the twelve programs with strong evidence, the business outcomes they connected to, and the Enhancement cycles that had improved them over time. She then described the fifteen programs in the middle category and the data collection she was building to evaluate them properly. She then named the sixteen and said what she intended to do about them: ten would receive focused evaluation over the next quarter to determine whether to invest or retire, and six she was recommending for retirement now. She had already spoken to the relevant business unit heads about the six.

The CHRO looked at the retirement list. "This is more honest than what I usually hear."

"I know," Claudette said. "We've had the wrong incentive. We've been tracking what we produce instead of what works. I'm trying to change that."

The CHRO approved the retirement list and asked two questions about the ten programs under evaluation. The meeting ended with fifteen minutes to spare and a specific commitment: the CHRO would connect Claudette directly with the three business unit leaders whose teams were generating the most L&D requests, so she could get ahead of next year's needs rather than responding to this year's emergencies.

It was not a perfect meeting. But it was the first one in three years that felt like a genuine conversation about what the function was actually doing, with a person who was genuinely interested in the answer.

Claudette went back to her desk. She opened the spreadsheet. She started marking programs.

Chapter Summary

- The L&D manager is responsible for the conditions that make quality possible across a team and portfolio, not just for the quality of individual programs. This requires influencing and enabling rather than doing, and the quality failures it prevents are often invisible until they have accumulated for months.
- Portfolio management in an AI-augmented environment means making the invisible visible: which programs are strategically aligned, which have quality evidence, which are current, which are consuming bandwidth they don't justify. AI accelerates the audit; the manager's judgment drives the decisions.
- Team capability development requires honest assessment of where the team currently sits on the AI maturity scale, and differentiated management for the resistor (clear standards, respectful expectation), the zealot (quality review is not optional), and everyone in between.
- The shared prompt library is the single most practical team capability investment an L&D manager can make: it accelerates onboarding, ensures quality consistency, and creates a compounding team knowledge asset. It requires governance to remain useful.
- The 90-day INDIE adoption roadmap moves through four phases: foundation (shared understanding of why), pilot (one project, full framework, real conditions), standard (shared language and first library entries), and scale (all projects, sustainable rhythm). The monthly community of practice session is what sustains adoption past initial enthusiasm.
- The curriculum coordinator's challenge is coherence and currency across a body of learning content. AI accelerates gap mapping, alignment verification, contradiction detection, and currency risk assessment.
- Organizational credibility for L&D comes from presenting evidence honestly, including programs without evidence, retirement decisions, and resource constraints, rather than curating only what looks good. AI gives managers the analytical capacity to produce this kind of account; the willingness to present it builds the trust that makes L&D a strategic partner.

Reflection Questions

For individual journaling or team discussion.

1. Apply the portfolio health matrix to your current program portfolio. Without looking at the data, estimate: how many programs would you say you have strong evidence for? Now look at the data. Was your estimate accurate? What does the gap between your estimate and the evidence tell you about how the portfolio is being managed?
2. Using the AI maturity assessment, honestly locate your team on the scale. Where is the distribution, who is at Level 1, who is at Level 3? What is the gap between your most advanced practitioner and your least advanced? What is that gap costing in quality consistency?
3. Does your team have a shared quality standard for AI-assisted work? If yes, how was it developed and how consistently is it applied? If no, what has prevented one from being established, and what would it take to build one in the next sixty days?
4. Claudette's shift, from reporting activity to demonstrating value, including the programs without evidence and the retirement list, required a kind of organizational courage. What in your current organizational context would make that kind of honest account difficult to present? What would make it easier?
5. The chapter argues that the monthly community of practice session is what sustains adoption past initial enthusiasm. Does your team have a structured mechanism for collective learning and knowledge sharing? If not, what is the smallest viable version you could introduce in the next month, and what would it require from you to sustain it?

Part Three · Chapter 15
The Learning Technologist, SME, and Adjacent Collaborators

There were four of them in the post-project retrospective, and each one had a different account of what had gone wrong. The instructional designer said the storyboard reviews had taken too long and the content had arrived too late. The eLearning developer said the storyboard she had finally received bore almost no relationship to the content the SME had actually approved. The project manager said the timeline had been unrealistic from the start and nobody had listened when she said so in week two. And Sinclair, the clinical subject matter expert who had given six hours of his time to three review cycles, said that he still didn't understand why the final module looked so different from the document he had reviewed, and whether the clinical information in it was actually accurate, because two of the scenarios didn't reflect how his team actually worked.

They were not wrong, any of them. They were describing the same project from four different vantage points, and from each vantage point the problem was real. What none of them had said, because it was uncomfortable, and because the retrospective was already tense, was that the four of them had never been in the same room at the same time during the project until this moment, that nobody had explained to Sinclair what the design process was or what his role in it was meant to accomplish, and that the learning technologist who managed the LMS deployment had flagged a completion tracking problem three weeks before launch that nobody had followed up on. The learning technologist, Peter, had not been invited to the retrospective.

The Collaborators That Make or Break the Work

Every chapter in Part Three so far has addressed a role that sits clearly inside the L&D function. This chapter addresses three roles whose relationship to L&D is more variable, sometimes embedded inside the function, sometimes adjacent to it, sometimes contracted in from outside, but whose contribution to the quality and success of learning programs is as significant as any of the roles addressed in previous chapters.

The learning technologist manages the systems infrastructure on which learning programs run: the LMS, the authoring tools, the analytics platforms, the xAPI data architecture, and increasingly the AI tools themselves. The subject matter expert provides the domain knowledge without which instructional design is structurally sound but content-wise hollow. The project manager holds the timeline, the budget, the dependencies, and the communication architecture that allows a team of specialists to produce a coherent outcome rather than a collection of individually competent but incompatible contributions.

What the retrospective above illustrates is what happens when these roles are not integrated into the INDIE workflow: they operate in parallel rather than in collaboration, each doing competent work in their domain while the seams between domains become the project's failure points. INDIE addresses this by treating all three as genuine participants in the design process, not as service providers who are called in at specific moments and otherwise left outside the room.

The Learning Technologist

The learning technologist's role is one of the most consequential and most undervalued in the L&D ecosystem. They are responsible for the systems infrastructure that determines whether learning programs can be delivered, tracked, analyzed, and improved, which means they are responsible for the operational foundation of the Enhancement phase, the technical feasibility of the Development phase, and the data quality on which every intelligent program decision depends. When the infrastructure works well, it is invisible. When it fails, when completion rates are inaccurate, when xAPI data is malformed, when the LMS cannot accept the SCORM package the developer built, the failure is highly visible, and it is often attributed to the program rather than the infrastructure.

In the INDIE framework, the learning technologist is not a post-development support function. They are a Design phase collaborator, an Implementation phase infrastructure builder, and an Enhancement phase data custodian. Their input in the IN phase, on what the LMS can track, what data formats are available, what delivery constraints the technical environment imposes, shapes design decisions that cannot be revisited cheaply once development has begun. Their absence from the IN phase is one of the most reliable predictors of expensive late-stage technical problems.

The Learning Technologist in Each INDIE Phase

In the IN phase, the learning technologist's primary contribution is constraint definition: what the technical environment can and cannot support. This includes LMS capabilities and limitations (completion tracking mechanisms, reporting granularity, xAPI vs. SCORM support, mobile delivery requirements), authoring tool constraints (interaction types available, file size limits, browser compatibility), and data architecture decisions (what learner behavior data will be captured, in what format, accessible to whom). These are design constraints, and they belong in the IN phase alongside objectives and curriculum architecture.

In the Development phase, the learning technologist's contribution shifts to technical review: ensuring that what is being built will function as intended in the target delivery environment. This is not a final QA step; it is an ongoing review that catches technical problems while revision is still inexpensive. A learning technologist who reviews a prototype in week three of development can catch an interaction design problem that would require substantial rework if discovered in week eight.

In the Enhancement phase, the learning technologist is the data infrastructure custodian: ensuring that the systems are configured to capture the data the Enhancement cycle requires, that the data quality is sufficient for AI synthesis tools to produce reliable analysis, and that the dashboard views described in Chapter 9 are technically feasible in the available platforms. Enhancement data quality is a technical problem before it is an analytical one, and the learning technologist is the practitioner who can address it.

Without INDIE	With INDIE
Technical constraints discovered at launch, when they are expensive to address	Technical constraints defined in IN phase as design inputs, when they shape decisions early
LMS configuration determined by what the developer built, not what the analysis required	LMS configuration designed alongside objectives to capture the specific data the Enhancement phase needs
Completion tracking inaccuracies discovered in Enhancement data after program has run	Completion tracking validated in a technical pilot before full rollout
xAPI and SCORM issues surface in stakeholder review when the developer is already on the next project	Technical review integrated into Development phase milestones, not added as a separate late-stage gate
AI analytics tools operating on incomplete or malformed data	Data architecture designed for AI synthesis from the start of the project
Learning technologist learns about program goals from the LMS configuration request	Learning technologist understands program goals, audience, and Enhancement metrics from the Phase-IN conversation

PROMPT: Technical Constraint Definition for INDIE IN Phase
I am an L&D designer beginning the IN phase for [PROGRAM TITLE]. I need to define technical constraints before finalizing the curriculum architecture. Please help me structure a conversation with our learning technologist to surface all relevant constraints. Known program parameters: Target audience and organizational context: [DESCRIBE] Anticipated delivery formats: [eLearning, ILT, blended, mobile, etc.] Enhancement data requirements: [DESCRIBE WHAT WE NEED TO MEASURE] Timeline: [DESCRIBE] Please generate a structured technical constraints interview guide for a 45-minute conversation with the learning technologist covering: 1. LMS capabilities and limitations relevant to this program (completion tracking, mobile support, xAPI vs SCORM, reporting granularity, user group configuration) 2. Authoring tool constraints for the planned delivery format (interaction types available, file size, browser/device compatibility, accessibility features, AI tool integration) 3. Data architecture for Enhancement (what learner behavior data can be captured, in what format,

accessible to which AI analytics tools, at what latency)

4. Integration requirements
   (HRIS data connections, single sign-on, performance system linkage, dashboard tool compatibility)

5. Known constraints from previous programs in this environment
   (recurring technical problems, workarounds currently in use, planned platform changes that affect this timeline)

Format as questions the designer asks, not a checklist the technologist fills out. The goal is a conversation, not a form.

Role Callout: Learning Technologist

The most important shift for learning technologists in an AI-augmented environment is from technical executor to technical advisor: moving upstream in the project lifecycle so that design decisions are made with full knowledge of what the technical environment can support, rather than discovering the gap after development is complete. This requires being in the Phase-IN conversation, not just the deployment meeting. The learning technologist who can articulate technical constraints in design language, "this LMS can track completion but not time-on-task, which means our Level 2 analysis will be limited to assessment scores," is the technologist who gets invited into the design process. The one who only speaks in platform terminology gets called when something breaks.

PROMPT: AI Governance and Tool Evaluation Framework

My L&D team is evaluating AI tools to integrate into our workflow.
I am the learning technologist responsible for the technical assessment.

Tools under consideration: [LIST TOOLS AND THEIR INTENDED USE]
Existing infrastructure: [LMS, authoring platform, analytics tools, data environment]
Organizational AI governance requirements: [DESCRIBE IF ANY]
Data privacy context: [DESCRIBE — industry, learner data sensitivity, relevant regulations: GDPR, FERPA, HIPAA, etc.]

Please produce a technical evaluation framework for each tool covering:

1. Integration compatibility
   Does it connect to our existing infrastructure? What are the technical requirements for integration? What breaks if it changes?

2. Data handling and privacy
   What learner data does it process? Where is that data stored?
   What are the data retention and deletion policies?
   Does it comply with our regulatory requirements?

3. Output quality and auditability
   Can outputs be traced back to their source data?
   Is there a human review step built into the tool's design?
   What failure modes are documented by the vendor?

4. Vendor stability and support
   How established is the vendor? What is the support model?
   What happens to our data if the vendor changes or closes?

5. Total cost of integration
   Not just licensing cost: implementation time, training required, ongoing maintenance, and the cost of switching if the tool proves inadequate

Produce a scoring rubric for each dimension and a recommendation format the design team can use to make the adoption decision.

The Subject Matter Expert

The subject matter expert is the most consequential external collaborator in most instructional design projects, and the one whose collaboration is most consistently mismanaged. Sinclair's frustration in the opening scene, six hours of review time, three cycles, and a final module that didn't reflect how his team actually worked, is not unusual. It is the norm in organizations where the SME collaboration model has not been designed deliberately.

The typical SME collaboration model runs something like this: the designer conducts an interview, the SME provides a document dump of everything they know about the topic, the designer builds something from that material, the SME reviews it for accuracy, provides extensive corrections that look like scope expansion to the designer but look like obvious fixes to the SME, and the cycle repeats until time runs out. The final product is usually accurate in the narrow sense, the information is correct, but it often misses the practical wisdom, the edge cases, and the contextual judgment that makes the difference between content that is technically right and content that actually helps someone perform better.

AI has changed the SME collaboration dynamic in ways that are not yet widely understood. On one hand, AI can reduce the volume of routine drafting that SMEs are asked to review, which should free their attention for the judgment-intensive content that actually requires their expertise. On the other hand, AI-generated content is often fluent and plausible enough that SMEs, who are reviewing for accuracy, not for pedagogical quality, may miss subtle errors that a domain novice would not know to look for. The blood pressure monitor with impossible numbers from Chapter 11 illustrates the failure mode. The solution is not less AI-generated content. It is a better-designed SME review process that directs expert attention to the places where expert attention matters most.

What SMEs Are Actually Good At

The mistake that most SME collaboration models make is treating the SME as a content source rather than as a professional whose expertise has several distinct components, each of which is valuable in a different way and at a different stage of the design process.

Declarative Knowledge
What is true about this domain: the facts, frameworks, procedures, standards, and conceptual models that define expert understanding. This is the content that most SME collaboration models focus on almost exclusively. **AI enables:** Generating first drafts of declarative content from documented sources; organizing and structuring known information; producing plain-language explanations of technical concepts **Human SME required for:** Verifying accuracy of AI-generated declarative content; identifying what is missing, what is subtly wrong, and what is technically correct but misleading without context

Procedural Knowledge
How skilled practitioners actually perform tasks: the steps, the decision points, the workarounds, the shortcuts that experience produces. This knowledge often exists in practice but has never been explicitly documented.

AI enables: Structuring and formatting documented procedures; generating decision trees from written protocols; producing job aid drafts from procedure documentation

Human SME required for: Describing how work actually happens (vs. how it is documented to happen); identifying the decision points that written procedures skip; narrating the expert reasoning that makes a procedure intelligible

Conditional Knowledge

When to apply which approach: the judgment that experienced practitioners develop about context, exceptions, edge cases, and the situations where the standard procedure doesn't apply. This is the knowledge that makes the difference between a competent novice and an effective practitioner.

AI enables: Generating scenario drafts that illustrate conditional knowledge application; producing first versions of decision frameworks; structuring case study outlines

Human SME required for: Identifying the conditions that trigger each approach; naming the edge cases that novices consistently miss; distinguishing between situations that look similar but require different responses

Cultural and Contextual Knowledge

How this domain functions in this specific organizational context: the norms, the politics, the unwritten rules, the relationships between teams, and the history that shapes how things are actually done here.

AI enables: Almost nothing. This is the most context-specific knowledge type and the one AI has the least access to.

Human SME required for: Everything. This knowledge only exists inside the organization, and only practitioners who have lived in the context can provide it. It is also the knowledge most frequently omitted from training content and most consequential for transfer.

Designing the SME Collaboration for INDIE

A well-designed SME collaboration in INDIE matches the type of SME contribution required to the appropriate phase of the project and structures the SME's time around what only they can provide rather than around reviewing drafts of things AI can produce more cheaply. This changes the SME experience from "reviewing a lot of content that is mostly correct but needs adjustment" to "contributing the knowledge that the program genuinely cannot be built without." The distinction is not merely motivational. It produces better programs.

The most important change is front-loading the conditional and contextual knowledge conversations. These conversations cannot happen from a review document. They require a different kind of session: the designer asking, "when does the standard procedure break down?" and "what does an expert know that a novice would miss?" and "can you walk me through a recent case where this went wrong?" AI can prepare the designer for these conversations, can help analyze the knowledge that emerges from them, and can produce the first drafts of instructional content that incorporates it. What AI cannot do is elicit this knowledge from the SME. That is the designer's job, and it requires time and skill that the typical content review meeting does not support.

PROMPT: SME Knowledge Elicitation Interview Design
I am preparing to interview [JOB TITLE / SME] about [TOPIC AREA] for a learning program targeting [AUDIENCE]. I have already gathered: [DESCRIBE AVAILABLE DOCUMENTS, PROCEDURES, EXISTING TRAINING CONTENT] AI has produced draft content for the declarative and procedural knowledge components. The gaps I need this SME to fill are in conditional and contextual knowledge: the judgment, the edge cases, and the organizational context that doesn't appear in documents. Please design a 60-minute knowledge elicitation interview that: 1. Opens by explaining the project purpose and the SME's specific contribution (5 min) — not a general overview of the program, but a specific account of what this conversation will produce that documents cannot 2. Elicits conditional knowledge (25 min): - What does this look like when it goes wrong, and what is the most common reason it goes wrong? - When does the standard procedure not apply, and how does an expert know? - What would a novice do that an experienced practitioner wouldn't, and why?

- Can you walk me through a recent case that was harder than it looked? What made it harder?

3. Elicits contextual knowledge (20 min):
   - How does this actually work here, as opposed to how it's documented to work?
   - Who in the organization is the expert others go to when they have a hard case in this area?
   - What has changed recently that someone new wouldn't know?

4. Closes with a validation question (10 min):
   - I'm going to show you the scenario I've drafted for this content. Does this reflect how this actually plays out? What's wrong with it?

Generate the specific questions for each section, not just the categories.

PROMPT: Preparing SMEs to Review AI-Generated Content

I need to brief a subject matter expert before they review AI-generated content for [PROGRAM TITLE]. The SME is [DESCRIBE: clinical expert, technical specialist, compliance officer, etc.] and has [DESCRIBE: extensive / limited] experience with training content review.

Content to be reviewed: [DESCRIBE scenarios, procedural content, assessment items, job aids]

Please draft a pre-review briefing for the SME that:

1. Explains in plain language how the content was produced (AI-generated draft, reviewed by designer, now needs domain accuracy review), without technical jargon or defensiveness
2. Tells them specifically what to look for, given that AI content is most likely to fail in these ways for this content type: [DESIGNER FILLS IN SPECIFIC RISKS, e.g., for clinical content: values on any depicted equipment, anatomical accuracy, protocol currency]
3. Tells them what NOT to spend time on:
   - Writing style and formatting (the designer's domain)
   - Whether the scenarios reflect best practice as they understand it (the designer needs to know what IS current practice first)
   - Complete rewrites (flag the issue, don't fix it)

4. Gives them a simple annotation format:
   FLAG: [element], [what's wrong], [what it should be]
5. Sets a realistic time expectation and expresses genuine appreciation for the specific contribution this review provides

Tone: Collegial, specific, respectful of their time.

Role Callout: Subject Matter Expert

If you are an SME who has been asked to contribute to a learning program, the single most useful thing you can do at the start of the project is ask the designer two questions: "What do you need from me that you can't get from a document?" and "When in the process is my judgment most valuable?" These questions reframe the collaboration from "SME reviews designer's work" to "designer and SME identify the contribution only the SME can make." The programs built through the second framing are better, and the SME's time is spent more productively. Many designers will not have thought to ask these questions of you. Asking them yourself changes the collaboration.

The Project Manager

Project management in instructional design projects is one of the most consistently undervalued contributions to program quality, and one of the most consistently under-resourced. Designers who are also managing their own projects, tracking timelines, managing stakeholder communications, coordinating SME schedules, monitoring budgets, and managing scope, are doing two cognitively demanding jobs simultaneously, and the design work usually degrades first because the project management pressure is more immediately visible.

When a dedicated project manager is present, they provide something qualitatively different from administrative support: they give the designer cognitive space to do design work. This is not a luxury. It is a quality factor, and its absence shows in programs built under the particular kind of cognitive fragmentation that project management overload produces.

AI has shifted the project manager's value proposition in an important way. The administrative and communication tasks that once consumed a significant portion of project management time, drafting status updates, generating timeline documents, producing meeting agendas, tracking action items, are all tasks that AI handles well. This frees the project manager for the work that AI cannot do: managing stakeholder relationships through ambiguity, making the judgment call about whether a scope change request is genuinely necessary or a symptom of unclear requirements, and being the person who says "this timeline is not realistic"' before the team has committed to it rather than after.

The Project Manager's AI Toolkit

The project manager's most valuable AI applications are in the administrative and communication work that consumes time without requiring the relational judgment that makes project management genuinely valuable. The following prompts are organized around the moments in a project lifecycle where AI provides the most meaningful time savings.

PROMPT: Project Kickoff Package
I need to prepare the project kickoff package for [PROGRAM TITLE], an L&D project with the following parameters: Project scope: [DESCRIBE] Team: [LIST ROLES — designer, developer, SME, facilitator, etc.] Stakeholders: [LIST WITH ROLES] Timeline: [START DATE to TARGET COMPLETION] Budget (if shareable): [DESCRIBE] Key risks already identified: [LIST] Deliverable milestones: [LIST MAJOR DELIVERABLES AND TARGET DATES] Please produce a kickoff package that includes: 1. Project charter (one page): Purpose, scope, success criteria, team roles and responsibilities, key milestones, and decision-making authority 2. RACI matrix for the six most critical project decisions: Who is Responsible, Accountable, Consulted, and Informed for each decision 3. Risk log with the five highest-probability risks: Probability (H/M/L), Impact (H/M/L), mitigation action,

and owner for each

4. Communication plan:
   Who receives what updates, at what frequency, in what format, and who owns each communication

5. Scope boundary statement:
   Three specific things this project WILL deliver and three specific things it will NOT deliver, stated clearly enough to manage scope creep conversations

PROMPT: Scope Change Assessment

I am managing [PROGRAM TITLE] and have received a scope change request.

Original scope: [DESCRIBE WHAT WAS AGREED]
Change requested: [DESCRIBE SPECIFICALLY]
Who is requesting it: [ROLE AND THEIR RELATIONSHIP TO THE PROJECT]
Reason given: [DESCRIBE]
Current project status: [TIMELINE POSITION, BUDGET CONSUMED, ETC.]

Please help me assess this scope change request by analyzing:

1. Genuine necessity vs. preference:
   Is this change addressing a gap that will cause the program to fail its objectives if not addressed? Or is it an improvement to something that already meets the objectives?

2. Impact assessment:
   Estimated time impact: [DESIGNER TO FILL IN, AI TO STRUCTURE]
   Estimated cost impact
   Dependencies affected (what else moves if this is approved)
   Quality impact if the original scope is maintained

3. Options:
   Option A: Approve as requested (full impact)
   Option B: Partial approval (what minimum version addresses the genuine need without the full scope expansion)
   Option C: Decline with rationale (when is this the right call)

4. Stakeholder communication draft:

A response to the requestor that acknowledges the request, presents the impact, and proposes the recommended option

Format this as a decision document I can present to the project sponsor, not just a memo to the requestor.

PROMPT: Timeline Recovery Plan

The project [PROGRAM TITLE] is behind schedule. I need to develop a recovery plan.

Current status:
Original target date: [DATE]
Current projected completion: [DATE]
Delay cause: [DESCRIBE — SME availability, content scope, technical issues, resource change, etc.]
Work remaining: [DESCRIBE MAJOR OUTSTANDING TASKS]
Resources available: [DESCRIBE — team, hours, budget]
Constraints: [DATE / DEPENDENCY / BUDGET THAT CANNOT MOVE]

Please produce a recovery plan that presents:

1. Three recovery options with tradeoffs:
   Option A: Compress — same scope, accelerated timeline (what does this require and what quality risks does it create?)
   Option B: Descope — reduced scope, original timeline (what comes out and what is the impact on program objectives?)
   Option C: Extend — original scope, later date (what is the organizational cost of the delay?)

2. A recommended option with rationale

3. A revised project timeline for the recommended option

4. A stakeholder communication that explains the situation, presents the options, and makes a specific recommendation without being defensive about how the delay occurred

Note: The stakeholder communication should take responsibility for the situation without assigning blame to specific team members.

Role Callout: Project Manager

The shift that AI enables for project managers on L&D projects is from administrative coordinator to strategic advisor on project health. When AI is handling status update drafting, meeting agenda preparation, and action item tracking, the project manager's time is freed for the work that actually determines project outcomes, the early conversation where a risk is named before it becomes a problem, the scope boundary negotiation that happens before requirements are locked rather than after, and the stakeholder relationship management that keeps a project aligned with organizational reality across its full lifecycle. The project manager who uses AI well becomes more valuable to the design team, not less.

Making Collaboration Work: The Integration Meeting

The structural change that would have prevented the retrospective in this chapter's opening scene is not a better communication plan or a more detailed RACI matrix. It is an integration meeting at the start of the IN phase, a session in which all the people who will contribute to the program are in the same room, understand each other's roles, and have a shared understanding of what they are building together before any of them begin their individual work.

This seems obvious. It almost never happens. The most common alternative is that each collaborator receives a briefing appropriate to their role at the moment they need to contribute, which means the learning technologist learns about the program when they receive the SCORM package, the SME learns about the design when they receive the storyboard, and the project manager learns about the real scope when the designer tells them the original timeline was unrealistic. Sinclair, who contributed six hours to a program he never fully understood, was in this position.

The integration meeting is not a long meeting. Sixty to ninety minutes, once, at the start of the project, with clear structure and a specific output. Its purpose is to give every collaborator the same foundational understanding: what the program is trying to accomplish, who the learners are, what each person's contribution to the outcome will be, and how their work connects to everyone else's. AI can prepare the agenda and synthesize the output. The presence of the people in the room is what makes it work.

PROMPT: Project Integration Meeting Agenda
I need to prepare an integration meeting agenda for the collaborators on [PROGRAM TITLE]. This is the first meeting where all contributors will be in the same room. Attendees: [LIST — include designer, developer, SME(s), learning technologist, project manager, and any other key collaborators] Duration: 75 minutes Program summary: [BRIEF DESCRIPTION OF PROGRAM, AUDIENCE, AND GOAL] Current project phase: [IN PHASE, BEGINNING OF PROJECT] Please design a 75-minute meeting agenda that: 1. Opens with a shared purpose statement (10 min): What are we building, who are we building it for, and what does success look like? Make this concrete enough that a person who knows nothing about instructional design can understand it. 2. Roles and contributions (20 min): For each person in the room, name: - What they will contribute that nobody else can - When in the project their contribution is most critical - What they need from each other person to do their work well 3. Technical environment overview (15 min): Learning technologist presents the key constraints and capabilities of the delivery environment. Designer asks the two most important constraint questions. 4. SME knowledge mapping (20 min): What does the SME know that isn't in any document? What are the edge cases, the judgment points, the contextual knowledge this program needs that only they can provide? 5. Risk and dependency identification (10 min): What are the three biggest risks to this project's success? What dependencies between team members most need to be managed? Produce the agenda with facilitation notes for each section, the specific questions the meeting needs to answer, and a template for the meeting output document.

The Retrospective, Reconvened

Twelve months after the retrospective that opened this chapter, the same four people were back in a room together, this time at the start of a new project. Peter, the learning technologist, had been invited.

The integration meeting had been the project manager's idea, proposed at the kickoff with the explicit framing that it was the thing they hadn't done the last time. It lasted seventy minutes. Sinclair, who had arrived with the slightly guarded posture of someone who remembered his six hours of review that had produced something he didn't recognize, left understanding what the instructional designer was trying to accomplish with the scenarios, why the document review he had done the last time had not served either of them well, and what the forty-five-minute knowledge elicitation session the following week was designed to produce.

Peter left understanding that the program needed xAPI tracking for scenario completion events, not just module completion, and that he needed to flag a configuration issue in the LMS that would prevent that without a workaround, a flag he had sent by email on the last project and that had been noted and not followed up on. This time he said it in the room. The designer wrote it down as a design constraint. The project manager added it to the risk log.

The project ran eight weeks. It was not without problems, the SME's knowledge elicitation session produced a conditional knowledge insight that required redesigning one of the three scenarios in week five, which was not in the original timeline. The project manager ran the scope change assessment, presented the options, and the sponsor approved a one-week extension. The change was made. The scenario was better for it. Sinclair, reviewing the final content, said it reflected how his team actually worked.

That sentence "it reflects how my team actually works" is the sentence that instructional design projects are trying to produce from their SMEs. It does not arrive from a document review. It arrives from a collaboration that was designed to make it possible.

Chapter Summary

- The learning technologist, subject matter expert, and project manager are not peripheral to INDIE; they are participants whose absence from the design process is one of the most reliable predictors of expensive late-stage problems. Their integration begins in the IN phase, not when they are needed to deliver a specific contribution.
- The learning technologist's most important INDIE contribution is upstream: defining technical constraints in the IN phase so they shape design decisions rather than limit them. Their Enhancement phase role as data infrastructure custodian determines the quality of every AI-assisted analysis the team performs.
- SME expertise has four distinct components: declarative knowledge (AI can draft), procedural knowledge (AI can structure), conditional knowledge (SME required), and contextual knowledge (SME exclusively). Designing SME collaboration to elicit the third and fourth components, rather than primarily reviewing the first and second, produces programs that reflect how work actually happens.
- AI has changed the SME review dynamic: AI-generated content is fluent enough that SMEs can miss subtle domain accuracy errors at normal review speed. A structured pre-review brief that directs SME attention to the specific failure modes of AI-generated content in this domain reduces post-launch corrections.
- The project manager's AI-augmented role shifts from administrative coordinator to strategic project health advisor. AI handles timeline documentation, status communications, and scope change analysis; the project manager provides the relational judgment and early-warning signals that determine whether those analyses are acted on.
- The integration meeting, one session, all collaborators, at the start of the IN phase, is the structural intervention that prevents the majority of collaboration failures in instructional design projects. It is almost never done and almost always worth doing.

Reflection Questions

For individual journaling or team discussion.

1. Think about the last project where collaboration between core L&D team members and adjacent collaborators broke down. At which seam did the breakdown happen, between design and technology, between design and SME, between design and project management? When in the project lifecycle did the breakdown become visible? When did the conditions for it actually form?
2. Consider your current or most recent SME relationship. Which of the four SME knowledge types did the collaboration successfully extract? Which types were left on the table? What would it have taken to access the conditional and contextual knowledge that the program may have needed?

3. Has your learning technologist ever flagged a technical constraint or problem that was not acted on until it caused a visible failure? What structural change to the project process would have surfaced that flag when it could still be addressed cheaply?
4. The chapter argues that the integration meeting is the structural intervention that prevents most collaboration failures in ID projects, and that it almost never happens. What prevents it in your current organizational context, time pressure, assumption that people already understand their roles, reluctance to commit to a meeting before work has begun? What would it cost, in time, to do it anyway?

Part Three · Chapter 16

The Solo Practitioner and Consultant

Lena had left her corporate L&D role fourteen months ago. She had done it for the right reasons: she was tired of the bureaucracy, she had wanted to work with a wider range of organizations, she had believed that her skills and reputation were worth more on the open market than her salary had reflected. All of that was probably true.

What she had not fully anticipated was the particular texture of working alone. Not the isolation: she had known about that and had prepared for it with a home office, a standing desk, and a subscription to a coworking space she used three days a week. What she hadn't anticipated was the cognitive weight of being every role simultaneously: the designer and the developer and the project manager and the quality reviewer and the business developer and the bookkeeper and the person who decided whether the proposal was ready to send or needed another pass.

The quality review problem bothered her most. In her corporate role, she had always had a colleague she could ask to read something before it went to a stakeholder. The colleague didn't have to be an expert; they just had to be another set of eyes that wasn't hers, that hadn't written the thing and couldn't see it as it was meant to be rather than as it was. She missed that colleague. She sent things out that she knew weren't quite right and that she would have caught if someone else had read them.

Six months into the practice, she started using an AI tool as that colleague. Not as a replacement for expertise, she understood its limitations clearly enough by then. As a different set of eyes. Something to read the thing as if it hadn't written it.

It was not the same as a colleague. But it was better than nothing, and in the ways that mattered most, catching the structural issue she had been too close to see, flagging the assumption the client wouldn't share, suggesting the third option she hadn't considered. It was often better than she expected.

The Solo Practitioner's Particular Situation

The solo instructional designer and the independent L&D consultant occupy a position in the field that is structurally different from every role addressed in the preceding chapters. They are not smaller versions of organizational practitioners. They are a different kind of practitioner entirely, operating under constraints and with freedoms that organizational roles do not share.

The constraints are real and specific. There is no team to distribute work across, no institutional knowledge to consult, no budget for specialized tools that get used once, no peer reviewer to catch what self-review misses, no administrative support to handle the non-design work that accumulates relentlessly. Every hour spent on project management, client communication, billing, and business development is an hour not spent on design. The cognitive context-switching between roles, designer to project manager to business developer to bookkeeper, sometimes within a single morning, creates a fragmentation that organizational practitioners are largely protected from by structure.

The freedoms are also real and specific. The solo practitioner can specialize in ways that organizational generalists cannot. They can work with a wider range of organizations than any single employer provides. They can set their own quality standards without navigating organizational politics about what "good enough" means. They can structure their day around when they do their best work. And they have, in many cases, a depth of accumulated cross-industry experience that organizational practitioners, who have spent years in one context, do not.

What AI has done for this population is not merely accelerate their work. It has changed the structural economics of solo practice in ways that are still being absorbed. A solo practitioner with AI tools can now offer analytical depth, production breadth, and quality consistency that previously required a team. The capacity gap between a solo practice and a small team has narrowed significantly. And the solo practitioner who uses AI well and understands its limitations clearly is offering something qualitatively better than the solo practitioner who uses it poorly or not at all, not because they are producing more output, but because they are able to bring more of the INDIE framework to bear on work that previously required the practitioner to choose which corners to cut.

The Solo Capacity Map

The table below maps each INDIE phase against the constraints solo practitioners have traditionally faced and the capabilities AI now provides. It is not a guarantee; AI output requires the same expert review in solo practice that it requires in team practice, and without a colleague to catch errors, the review discipline is more important, not less. It is an honest account of what has changed.

INDIE Phase	Without AI: Solo constraint	With AI: Solo capability
IN: Investigation-Design	Needs analysis often compressed to a single stakeholder conversation; persona development done intuitively rather than systematically; objective writing and curriculum architecture done in a single sitting under time pressure	Multi-source data synthesis completed rapidly; persona drafts generated from available data and refined with stakeholder input; objective quality review and curriculum optimization available without a second reviewer
D: Development	Content drafting, scenario writing, assessment creation, and storyboarding all done sequentially by one person; quality review is self-review, which has known limitations; cognitive load audit requires mental distance that is hard to achieve from your own work	AI generates first drafts of scripts, scenarios, assessment items, and facilitation guides; the solo practitioner's expertise goes into evaluation and revision rather than initial production; cognitive load audit available through AI perspective prompts
I: Implementation	Facilitation preparation done alone; no colleague to debrief with after delivery; stakeholder communications drafted in compressed time between project deliverables	Preparation scenario generation, facilitation guide internalization, and post-session synthesis all accelerated; stakeholder communication drafts available within minutes; the solo practitioner

		can prepare as thoroughly as a team
E: Enhancement	Evaluation data often goes unanalyzed because there is no capacity left after delivery; Level 3 and 4 measurement rarely attempted because the infrastructure cost is prohibitive for a solo practice	Feedback synthesis automated; pattern analysis available from small datasets; Enhancement reports drafted quickly and calibrated to different audiences; Level 3 follow-up structured and managed with AI support
Business Development	Proposals written from scratch for each client; case studies and capability statements updated infrequently because updating is low-priority compared to billable work	Proposal frameworks maintained and adapted quickly; capability statements updated incrementally; case study drafts generated from project notes; the solo practitioner can respond to RFPs at a speed that was previously impractical

AI as the Missing Colleague

Lena's discovery: that AI could function as a different set of eyes on work she was too close to evaluate clearly, is the most underappreciated AI application for solo practitioners and one that has no direct equivalent in team practice. When you have colleagues, the "second reader" function is available as a social act: you send something and ask someone to take a look. When you work alone, that function is either absent or self-performed, and self-review has a well-documented limitation: the writer reads what they intended to write, not what is actually on the page.

AI addresses this specific gap with specific effectiveness. An AI tool asked to evaluate a design document does not know what you intended to say. It reads what is there. It will identify the assumption that you made without stating, the step in the sequence that is missing, the objective that doesn't actually meet Mager's criteria despite looking like it does, the scenario that would confuse a learner who doesn't have the background context you have. These are exactly the things that a colleague would catch. And they are exactly the things that solo review misses.

This function works best when the prompt explicitly assigns the AI a critical perspective rather than an evaluative one. "Review this for quality" produces a different, less useful response than "read this as a skeptical client seeing it for the first time and tell me what you would question." The AI's perspective can be assigned. Assigning it deliberately is what makes the function useful rather than merely validating.

PROMPT: The Missing Colleague Review
I need a critical review of the following [DELIVERABLE TYPE: design document / learning objectives / storyboard section / proposal / etc.] before it goes to [CLIENT / STAKEHOLDER / PRODUCTION]. [PASTE CONTENT TO REVIEW] Context: This is for: [DESCRIBE CLIENT AND THEIR CONTEXT] Audience for this deliverable: [WHO WILL READ / ACT ON THIS] What I'm most uncertain about: [NAME YOUR OWN DOUBT IF YOU HAVE ONE] Please read this as a skeptical, intelligent [CLIENT / COLLEAGUE / LEARNER] encountering it for the first time. Do not read it charitably. Tell me: 1. What assumption did I make that I didn't state? 2. What is the weakest part of this, and what specifically makes it weak? 3. What question would this raise for the intended audience that I haven't answered? 4. What would you change if this were yours? 5. Is there anything here that is technically correct but would not land well with this specific audience or in this specific context? Do not tell me what is good about it. I can evaluate that myself. Tell me what to fix.

Client Management Without a Support Structure

The solo practitioner's client management work, scoping projects, writing proposals, managing expectations, handling scope creep, delivering difficult feedback, maintaining relationships across a portfolio of clients simultaneously, is work that organizational practitioners have administrative and managerial support for. The solo practitioner does it alone, usually between deliverables, often in compressed time, and always with the awareness that every client relationship is also a source of future work or a reference.

This is the work that most solo practitioners find most draining and most difficult to do well. It is also the work where the stakes of doing it poorly are highest: a scope mismanagement problem on an organizational team is a project problem. On a solo practice, it can be a reputation problem.

Scoping and Proposals

The proposal is the solo practitioner's most important deliverable, because it defines the scope before work begins and provides the reference point for every scope conversation that follows. Proposals that are vague about what is included are proposals that invite scope creep. Proposals that are specific about what is not included are proposals that protect the practitioner's time and the client's expectations simultaneously.

AI can draft proposals quickly from a set of project parameters, and the drafts are often structurally sound. The practitioner's contribution is the commercial and relational judgment that the AI does not have: whether the proposed fee reflects the actual time required, whether the scope boundaries are stated specifically enough to hold in a difficult conversation, and whether the proposal's tone reflects the relationship accurately.

PROMPT: Project Scoping and Proposal Draft
I need to write a proposal for an instructional design project. Client context: Organization: [DESCRIBE size, industry, L&D maturity] My relationship with this client: [NEW / EXISTING, DESCRIBE] Who will receive this proposal: [ROLE AND SENIORITY] Project parameters: Request: [DESCRIBE WHAT WAS ASKED FOR] My assessment of what's actually needed: [DESCRIBE IF DIFFERENT] Scope as I currently understand it: [DESCRIBE] What is explicitly NOT included: [LIST] Timeline: [DESCRIBE CONSTRAINTS AND PROPOSED DATES] Deliverables: [LIST WHAT WILL BE PRODUCED] My proposed fee structure: [DESCRIBE fixed / day rate / phased] Please draft a proposal that: 1. Opens with a brief problem statement that demonstrates I understand their situation (not just their request) 2. Proposes a clear approach organized by INDIE phase (without using the word INDIE unless the client knows it) 3. States deliverables specifically, including file formats, review cycles, and what "done" means for each 4. States scope boundaries explicitly in plain language (what is NOT included and what triggers a scope conversation) 5. Presents the investment clearly with the payment structure 6. Closes with a specific next step Tone: Professional, direct, warm. Treats the client as an intelligent adult who can handle specificity. Does not over-promise. After the draft, flag the two scope boundary statements that are most likely to need strengthening based on the project description.

Managing Scope Creep Alone

Scope creep is the solo practitioner's most reliable source of unpaid work and eroded client relationships. It arrives in forms that are individually small and collectively significant: one more revision cycle that wasn't in the proposal, a new module that emerged from a stakeholder conversation the client didn't think to flag, a request to facilitate the program in addition to designing it, a "quick" addition that is not quick. Each instance feels manageable in isolation. The cumulative effect is a project that has expanded by thirty percent without a corresponding expansion of the fee.

The solo practitioner who manages scope well does not do so by being rigid about the original proposal. They do so by having the scope conversation early, explicitly, and without apology: "That's outside the original scope. Here is what it would add to the timeline and the investment. Do you want to proceed?" This is a business conversation, not a negotiation, and treating it as such, matter of fact, not defensive, is the habit that separates practitioners who are sustainably profitable from those who are perpetually overworked.

PROMPT: Scope Conversation Preparation
I need to have a scope conversation with a client about work that has expanded beyond the original proposal. Original scope: [DESCRIBE WHAT WAS AGREED] What has been requested beyond scope: [DESCRIBE SPECIFICALLY] My estimate of the additional effort: [HOURS / DAYS] The relationship context: [NEW CLIENT / LONG-TERM CLIENT / SENSITIVE MOMENT] What I'm most concerned about in this conversation: [NAME IT] Please help me prepare by: 1. Drafting the opening sentence of the conversation: Clear, direct, non-apologetic, relationship-preserving 2. Anticipating the client's three most likely responses and drafting a response to each that: - Maintains the scope boundary without being adversarial - Gives the client a genuine choice (approve the addition, descope something else, decline the addition) - Preserves the relationship regardless of what they decide 3. Drafting a written follow-up that documents the outcome of the conversation and confirms what was agreed

(whether that is an approved addition, a declined addition,
or a scope trade)

Note: The goal is a fair outcome for both parties, not winning
the negotiation. A client who feels managed is not a long-term client.

Delivering Difficult Feedback

The solo consultant's client relationship requires navigating feedback in both directions: receiving feedback from clients on work the practitioner is proud of, and delivering feedback to clients on requests or decisions that the practitioner believes are wrong. The second is harder. It requires the professional confidence to say "I don't think this is the right approach, and here is why" to someone who is paying for the practitioner's expertise and may not be expecting to hear that their idea is a problem.

The parallel to Chapter 13's courage competency is direct: the solo consultant who tells a client what they want to hear rather than what the evidence shows is not serving the client's interests. They are protecting the immediate relationship at the cost of the long-term one because clients who receive poor advice and produce poor outcomes eventually attribute the outcome to the consultant who endorsed the approach.

PROMPT: Delivering Difficult Professional Feedback to a Client

I need to deliver feedback to a client about a decision or direction
I believe is professionally problematic.

The situation: [DESCRIBE — what they want to do, why I think it
is the wrong approach, what the risk is]
My evidence: [DESCRIBE WHAT SUPPORTS MY CONCERN]
The client relationship: [DESCRIBE — tenure, rapport, their likely
reaction to being pushed back on]
What I'm most worried about saying: [NAME THE SENTENCE YOU'RE
AVOIDING]

Please help me prepare this feedback by:

1. Drafting the feedback in three versions:
   Direct: States the concern clearly, immediately, without softening
   Contextual: Opens with what I understand about their reasoning,

then names the concern
Question-led: Opens with a question that surfaces the risk rather than naming it

2. Recommending which version fits this client and this moment, and why

3. Drafting the "what I recommend instead" section:
The feedback is only useful if I have a specific alternative.
Help me articulate it clearly.

4. Identifying what I should be prepared for:
The response that means the client is genuinely hearing this, the response that means they're not, and what I do in each case

Note: The goal is to serve their interests, not to be right. Frame the feedback that way.

Quality Without a Peer

The quality review challenge that Lena encountered is real for every solo practitioner, and it is not fully solved by AI. What AI provides is a partial substitute for a specific component of peer review: the external perspective that catches what self-review misses. What it cannot provide is the professional expertise to evaluate whether the design is pedagogically sound, whether the objectives are the right ones, whether the scenario reflects the learner's actual situation rather than a plausible version of it. Those evaluations require the practitioner's own judgment applied rigorously, without the benefit of another expert looking over their shoulder. Three practices protect quality in solo work in ways that AI alone cannot replicate.

- **Time distance.** The most reliable quality review technique for solo practitioners is temporal: complete a draft, put it away for at least twenty-four hours, then read it as if encountering it for the first time. The errors that are invisible immediately after writing become visible after cognitive distance has been established. AI can be used in the gap to generate parallel work, but the wait is non-negotiable. Build it into project timelines explicitly, a review day between draft completion and client delivery is not a luxury, it is a quality standard.

- **Professional community.** The solo practitioner who participates actively in a professional community, an industry association, an online L&D community, a peer group of other consultants, has access to the collective judgment that their daily work does not provide. This is not primarily about networking. It is about having peers who can look at a design decision and tell you whether they have encountered the same problem and what they did about it. The solo practitioner who is isolated from professional community produces work that reflects only their own accumulated experience, which is a meaningful limitation.
- **Structured self-critique.** Before any significant deliverable goes to a client, run it through the relevant INDIE quality checklist explicitly and in writing, not mentally. A mental QA passes things that a written QA catches, because writing forces precision that thinking permits to be vague. The Development phase QA checklist in Chapter 7, the Implementation Readiness Test in Chapter 7, and the objective quality review prompt in Chapter 6 are all tools designed for this purpose. Use them on your own work before a client does.

Role Callout: Solo Practitioner and Consultant

The most important structural investment a solo practitioner can make in an AI-augmented environment is not in a broader tool suite. It is in the discipline of treating AI output as a first draft that requires expert review, not a finished product that requires only formatting. The solo practitioner's competitive advantage over less experienced practitioners who also have AI tools is not access to the tools. It is the expertise to evaluate their output accurately and the professional judgment to know when the output is wrong. That advantage is only present if the review is genuinely rigorous. A solo practitioner who accepts AI output without substantive review is not more capable than they were before AI. They are faster at producing work of uncertain quality, which is not a competitive advantage and is a reputational risk.

Building a Solo Practice on a Realistic Budget

One of the most common misconceptions about AI-augmented solo practice is that it requires significant tool investment. It does not. The most important AI applications for instructional design, drafting, synthesis, review, and analysis, are available at the free or low-cost tier of the major AI platforms. The marginal return on premium tool subscriptions is real but incremental. A practitioner building a practice on a tight budget can implement the majority of the INDIE workflow with minimal tool expenditure if they are willing to invest in learning to prompt well.

The toolkit guide below is organized by budget tier. It is not a product endorsement, specific tools change faster than book publication cycles, and the category of tool matters more than the specific product. It is an honest account of what is required at each level of practice to implement INDIE effectively.

Minimal Budget
• AI writing assistant: Free tier of a major LLM (sufficient for most prompting tasks with thoughtful prompt design)
• eLearning authoring: Articulate Rise (subscription) or free alternatives like iSpring Suite demo / Canva for simple modules
• Transcription and captions: Whisper (open source) or Otter.ai free tier
• Project management: Notion free tier or Trello
• Portfolio and proposals: Canva free tier for visual assets
At this tier, the primary investment is time: learning to write prompts that compensate for tool limitations through specificity and iteration. Most of the INDIE workflow is achievable.

Working Budget
• AI writing assistant: Paid subscription to a premium LLM (faster, higher context window, better reasoning on complex prompts)
• AI image generation: Midjourney or similar (replaces stock photo subscription for many use cases)
• AI voice synthesis: ElevenLabs or Murf starter plan (for narration on appropriate content types)

• eLearning authoring: Articulate 360 (Storyline + Rise + Review for client collaboration)
• Video: Descript for screen recording, transcription, and basic editing
• Analytics: Google Analytics or LRS (Learning Record Store) for xAPI if clients need it
At this tier, the full INDIE production workflow is viable for most client engagements. The combination of AI writing, image, and voice tools replaces what a team would previously have required.

Established Practice
• AI writing assistant: Premium plan with API access for custom workflow integration
• Specialized AI tools: Domain-specific tools if your practice has a specialty (e.g., healthcare simulation, compliance automation)
• LMS access: Hosted LMS for client delivery (or reseller partnership with a platform vendor)
• Enhanced production: Full Adobe Creative Cloud or equivalent for high-end visual production
• Business tools: CRM, proposal software (Proposify or similar), time tracking, accounting
• Continued learning: Conference attendance, community memberships, specialized courses
At this tier, the practice has a defined market position and the tools serve that position. The investment is in specialization and business infrastructure, not general capability.

A note on free tiers: the major AI platforms' free tiers are more capable than most practitioners realize. The primary limitations are context window size (the amount of content you can include in a single prompt) and response speed, not output quality on most instructional design tasks. A practitioner who learns to structure prompts for smaller context windows can accomplish most of the INDIE workflow at zero tool cost. This is worth knowing before spending money on subscriptions.

The Specialization Advantage

The solo practitioner who has not specialized is competing on generalist terms with organizational L&D teams, which is an unfavorable comparison on most dimensions: teams have more capacity, more institutional knowledge, more relationship infrastructure, and more organizational reach. The solo practitioner who has specialized, who is specifically excellent in a domain, an industry, a modality, or a type of problem, is not in that competition. They are offering something that most organizational teams cannot provide: depth of specialized expertise applied with full professional attention to the client's specific situation.

AI strengthens the specialization advantage in a specific way: it handles the generalist production work that once consumed the solo practitioner's time, freeing capacity for the specialized judgment that represents their actual market value. A compliance training specialist whose AI tools handle first-draft scripting, scenario generation, and assessment item production can direct their attention to the regulatory precision, the domain accuracy, and the organizational context sensitivity that makes their work worth paying a specialist premium for. A leadership development consultant whose AI handles the research synthesis and content drafting can invest in the diagnostic conversations, the coaching relationships, and the organizational systems thinking that distinguishes excellent leadership development from adequate leadership development.

The practical implication: solo practitioners should actively resist the temptation to use AI's expanded capacity for generalist work to take on a wider variety of engagements. The capacity gain is best invested in doing the specialized work more deeply, not in doing more different kinds of work.

PROMPT: Specialization Positioning Statement
I am developing the positioning for my solo instructional design practice. I want to articulate my specialization clearly enough that ideal clients recognize themselves in it. My background: [DESCRIBE — years of experience, industry context, types of work, notable engagements] What I am genuinely better at than most ID practitioners: [NAME IT, even if it feels immodest] The clients I do my best work with: [DESCRIBE — size, industry, L&D maturity, the type of problem they bring me] The work I find most energizing: [DESCRIBE] The work I could do but that others could do just as well: [NAME IT] Please help me develop a positioning statement that: 1. Names my specialization in terms of client outcomes, not my credentials or methods ("I help [WHO] achieve [WHAT] by [HOW]" is a useful structure) 2. Is specific enough that ideal clients say "that's exactly what I need and non-ideal clients can self-select out 3. Does not use generic L&D language ("learning solutions," "performance improvement," "blended approaches") that every practitioner could claim 4. Suggests the AI-augmented depth I can bring to this specialization without making AI the centerpiece of the positioning (clients buy expertise, not tools) Draft three versions: one that leads with industry, one that leads with problem type, one that leads with outcome. I'll choose the one that most accurately reflects how I want to be known.

On Professional Isolation and the Social Construction of Knowledge

Vygotsky's insight that knowledge is constructed socially, that thinking develops through interaction with others, not in isolation, applies to the solo practitioner's professional development as directly as it applies to the learners they design for. The solo practitioner who works in professional isolation does not merely miss social connection. They miss the intellectual friction, the challenge, the alternative perspective, the question that reframes a problem, that collaboration provides. AI is a partial substitute for this friction when prompted to provide it (the "missing colleague review" is exactly this). But it is a partial substitute, not a complete one. Professional community provides forms of cognitive challenge, professional accountability, and collaborative knowledge construction that AI cannot replicate. The solo practitioner who actively maintains professional community is not just staying connected. They are maintaining the intellectual conditions that make their work continue to develop.

Sustainable Practice

The economic sustainability of a solo practice in an AI-augmented environment depends on a shift that many practitioners have not yet made: pricing expertise rather than time. When AI compresses the time required to produce a deliverable, the practitioner who charges by the hour earns less for the same quality of outcome. The practitioner who charges by the value of the outcome, by what the work is worth to the client, not by how long it takes to produce, benefits from the efficiency gain rather than being punished for it.

This requires a different client conversation about fees. It requires the practitioner to be able to articulate clearly what their expertise is worth, not the hours of work the client will see, but the accumulated knowledge, judgment, and experience that produces outcomes the client cannot produce without them. That conversation is harder than presenting a day rate, but it is more honest about what is actually being purchased.

AI also changes the sustainable capacity ceiling for solo practice. A practitioner who was productively at capacity managing three concurrent clients may find they can manage four or five with AI handling the production work. This is only advantageous if the additional capacity is deployed into quality, not into volume. The practitioner who takes on five clients and delivers work at the depth that three clients previously received has expanded sustainably. The practitioner who takes on five clients and delivers work at reduced depth to each has expanded unsustainably, and the Enhancement data, in the form of client satisfaction and repeat engagement, will eventually show it.

PROMPT: Practice Health Check
I want to assess the health of my solo practice and identify where to invest in the next quarter. Current practice overview: Active clients: [NUMBER AND BRIEF DESCRIPTION OF EACH] Revenue in last 90 days: [APPROXIMATE — OR JUST DESCRIBE AS ABOVE / AT / BELOW TARGET] Work that energized me most: [DESCRIBE] Work that drained me most: [DESCRIBE] Work I turned down: [DESCRIBE IF ANY] Work I wish I'd turned down: [DESCRIBE IF ANY] Tools and workflows I added: [DESCRIBE] Tools and workflows that didn't work: [DESCRIBE] Professional development: New capabilities I developed: [DESCRIBE] Skills I want to build: [DESCRIBE] Professional community I'm part of: [DESCRIBE] Please produce a practice health summary that: 1. Identifies the pattern in what kind of work and which clients are producing the best outcomes (for them AND for me) 2. Names the areas where my practice is misaligned with my stated specialization 3. Proposes three specific investments for the next quarter: one in capability, one in client relationships, one in the practice structure itself 4. Identifies one thing I should stop doing Frame this as a strategic review, not a performance assessment. The goal is direction, not evaluation.

Lena, Fourteen Months In

Fourteen months after leaving her corporate role, Lena had three active clients, a fourth in proposal stage, and a clearer sense of her market position than she had ever had in an organizational role where her work had been defined by whoever requested it.

She had settled into a practice focused on healthcare and life sciences organizations that were building learning programs for clinical and regulatory staff. It was demanding work, the domain accuracy requirements were higher than any other context she had worked in, the SME relationships were complex, and the compliance implications of errors were significant. It was also the work she was best at, and the work that her accumulated experience in the sector made her genuinely better at than most instructional designers who had not spent years in it.

Her AI toolkit had stabilized around three tools: a premium LLM for drafting and analysis, a voice synthesis tool for procedural narration on content that didn't require the emotional register of a human voice, and a transcription tool she ran on every stakeholder interview before her synthesis sessions. She had built a prompt library of forty-three entries, organized by INDIE phase, that she updated every time she found something better than what she had. She reviewed AI output against a domain-specific QA checklist she had built from the failure modes that clinical content was most susceptible to.

The missing colleague problem was not fully solved. There were still moments when she wanted a person rather than a prompt, when the question was not "what is wrong with this?" but "am I thinking about this correctly?" For those moments, she had cultivated two relationships: a former colleague who was now also consulting, with whom she had a standing bi-weekly call, and a mentor she had met at a professional association conference who had agreed to a monthly hour in exchange for occasional assistance with technology questions. Neither relationship was formal. Both were professionally essential.

At the end of the fourteenth month, she ran the practice health check prompt on the quarter's data. The output confirmed what she already knew but had been avoiding: one of her three active clients was consuming forty percent of her time and generating twenty percent of her revenue. The work was not bad. It was just not her best work, and it was not where her specialization made a genuine difference.

She sent a polite, professional message to that client the following week, proposing to complete the current engagement and not extend the relationship. She had the AI draft the message three different ways before she chose the one that was honest without being unkind. She revised it substantially before sending it.

The client was surprised but gracious. The space the engagement had occupied was filled within six weeks by a new project from a referral in her specialty. The new project was harder, more demanding, and exactly the kind of work she had left her corporate role to do.

This is what sustainable solo practice looks like in practice: not the absence of difficult decisions, but the clarity to make them because the work is aligned with what you are genuinely best at and what you genuinely want to offer. AI gives you the capacity to sustain that alignment over time. The clarity about what the alignment should be is yours.

Chapter Summary

- The solo practitioner's relationship with AI is qualitatively different from the organizational practitioner's: AI doesn't just accelerate their work; it changes what one person can offer. The capacity gap between a solo practice and a small team has narrowed significantly.
- AI's most distinctive application for solo practitioners is the "missing colleague review:" an explicitly critical perspective prompt that provides the external viewpoint that self-review cannot. The function works best when the prompt assigns AI a skeptical perspective rather than an evaluative one.
- Client management without organizational support requires specific AI applications: proposal drafting with explicit scope boundaries, scope conversation preparation, and difficult feedback structuring. In each case, the practitioner's relational judgment shapes the final communication; AI handles the structural drafting.
- Quality without a peer requires three practices that AI alone cannot replace: time distance (twenty-four hours between draft and review), professional community (the intellectual friction that isolation removes), and structured self-critique (written QA, not mental).
- Building a solo practice on a realistic budget is feasible: the majority of the INDIE workflow is achievable at the free or low-cost tool tier. The primary investment is in learning to prompt well, not in tool subscriptions.
- The specialization advantage is amplified by AI: the capacity freed from generalist production work should be invested in specialized depth, not in broader engagement variety. Solo practitioners should resist the temptation to expand into adjacent work just because AI has made them faster.

- Sustainable practice requires pricing expertise rather than time; a conversation AI enables by handling the production work while the practitioner's expertise creates the value. The quarterly practice health check maintains strategic alignment when no manager or organizational structure provides it.

Reflection Questions

For individual journaling or team discussion.

1. Lena's quality review problem, sending things out that she knew weren't quite right because there was no colleague to catch what she missed, is a specific and common solo practice challenge. How do you currently solve this problem? Have you tried using AI as a missing colleague reviewer? What was useful and what was not?
1. Run the missing colleague review prompt on a recent deliverable you were satisfied with before sending it. What did it find that you had missed? What does that tell you about where your self-review has consistent blind spots?
2. If you are currently in solo practice, apply the solo capacity map to your last three months. In which INDIE phases are you genuinely using AI to its potential, and in which are you still operating as if you were before AI tools existed? What would change if you closed the gap in the phases where you're under-utilizing AI?
3. The chapter argues that the capacity gain from AI should be invested in specialized depth, not broader engagement variety. Is your current practice moving toward greater specialization or toward greater breadth? What is driving that direction, and is it a deliberate strategic choice or an accumulated series of individual client decisions?

Part Four · Chapter 17

AI Governance, Ethics, and the Designer's Responsibility

The training module had been running for four months before anyone noticed. It was a customer service onboarding program, built for a retail organization with approximately twelve hundred customer-facing employees. The design had used AI-generated scenarios throughout: customer interactions, de-escalation situations, policy conversations. The scenarios were reviewed for accuracy. They were reviewed for cognitive load. They were reviewed for alignment with the learning objectives. Nobody reviewed them for who was in them.

What the review had not caught was a pattern that had emerged from the AI generation process: in the scenarios depicting difficult customer interactions, the ones involving complaints, escalations, and policy disputes, the customers were consistently portrayed with names and contextual details that coded them as non-white. The pleasant, cooperative customer scenarios had a different demographic distribution. The pattern was not explicit. It was not deliberate. It was a reflection of patterns in the AI's training data, reproduced without awareness into learning content that was now shaping how twelve hundred employees thought about difficult customer interactions.

The person who noticed was a new hire in her third week of employment, in a feedback form she submitted at the end of her onboarding. She named what she saw directly: "The difficult customers in these scenarios all seem to look the same. I don't think that's an accident."

It was not an accident. It was an oversight. The distinction is important, and so is this: in the impact it had on learners, the distinction is irrelevant. What the employees learned, over four months of onboarding content, was a pattern about who difficult customers tend to be. The training had taught it. Nobody had meant to teach it.

The Ethical Stakes of AI in Learning

Every chapter in this book so far has been primarily about capability: what AI enables, how to use it well, where its limits lie. This chapter is primarily about responsibility: what instructional designers are accountable for when AI tools are part of their workflow, what can go wrong in ways that cause real harm to real people, and what systematic practices prevent that harm.

The ethical stakes of AI in learning are not hypothetical. They are active and consequential. Learning content shapes how people think, about their work, about the people they work with, about the populations they serve. When that content is produced with AI tools that encode and reproduce biases, that present inaccurate information with the fluency of accurate information, that generate assessments measuring the wrong things, or that use learner data in ways that were never disclosed or consented to, the harm is not abstract. It is measurable in the performance of individual employees, the fairness of organizational decisions, and the dignity of learners who encounter their own communities represented in demeaning or distorted ways.

Instructional designers are generally not ethicists. They are not data privacy lawyers. They are not statisticians capable of auditing AI. What they are is the practitioners who make the decisions that determine what learners encounter, which scenarios get built, which assessments get deployed, which data gets collected, which AI outputs get reviewed and which get accepted without question. Those decisions have ethical dimensions, and the INDIE framework requires practitioners to engage with those dimensions deliberately, not accidentally.

> *The learning content practitioners create shapes how people think. That is exactly what makes it powerful. It is also exactly what makes it consequential.*

The Five Ethical Domains

The ethical landscape of AI-augmented instructional design organizes around five domains. Each is distinct in its nature and its practical implications. Each is addressable through specific design and governance practices. None is addressable through good intentions alone.

1. Bias and Representation

AI systems learn from human-generated data, and human-generated data reflects the biases, assumptions, and omissions of the humans who produced it. This is not a flaw that will be corrected as AI systems improve; it is structural. The data that trains AI systems is itself a product of societies that have majority and minority populations. It is not at all surprising that the patterns of those majority populations turn up in the data more often than that of the minorities; the patterns that AI learns to reproduce are statistical reflections.

In instructional design, bias in AI output manifests in four primary ways:

- **Demographic defaults in scenario generation.** When prompted without explicit demographic specification, AI scenario generation defaults to a narrow demographic range for characters, settings, and situations. The customer service case at the opening of this chapter illustrates one dimension of this. Other dimensions include: leadership scenarios that default to male characters, professional scenarios that default to white-collar settings, family scenarios that default to nuclear family structures, and communication scenarios that default to a single cultural register.
- **Uneven competence attribution in examples.** AI-generated examples and case studies tend to attribute expertise, leadership, and success to characters whose demographic presentation reflects the groups in the contexts the AI was trained on. In healthcare training, this can mean physician characters are consistently male and nurse characters are consistently female. In business training, characters may default to a demographic narrow enough to exclude most of the actual learner population.

- **Language register assumptions.** AI writing tools calibrate to a default professional register that reflects a specific educational and cultural background. Content generated at this register may be inaccessible to learners whose primary language is not English, whose educational background differs from the default. The AI does not flag this as a register choice. It presents it as simply what professional communication sounds like.
- **Feedback and assessment calibration.** AI-generated assessment items and feedback language may implicitly reflect cultural assumptions about what correct performance looks like that do not match the experience of all learners. A scenario built around a particular cultural approach to conflict resolution may mark a learner's culturally appropriate response as incorrect because it doesn't match the AI's training data default.

> **Case Study: The Pattern That Taught Without Intending To**
>
> The scenario described at the opening of this chapter is a composite, but it reflects a pattern that L&D practitioners have reported encountering in AI-generated content across multiple sectors. The mechanism is consistent: AI generates scenarios based on statistical patterns in training data; those patterns reflect historical associations that are themselves products of bias; the resulting content encodes those associations into training material without any human making an explicit decision to do so. The risk is greatest in exactly the scenarios where representation matters most: the difficult conversation, the policy enforcement, the performance management discussion, the safety incident.

The practical response to bias risk is not to avoid AI scenario generation. It is to consider demographic specifications when prompts are being engineered. Scenario prompts should specify the demographic of characters explicitly when it is appropriate to do so. This is a very important consideration when training is begin created for an international audience.

2. Accuracy and Intellectual Integrity

Chapter 2 addressed AI's propensity to confabulate, to produce fluent, confident, plausible-sounding content that is factually incorrect. In the context of governance and ethics, this is not merely a quality problem. It is an integrity problem.

Learning content that contains inaccurate information, whether the inaccuracy is a fabricated research citation, a misrepresented regulatory requirement, an outdated clinical protocol, or a subtly incorrect technical specification, does not merely fail to teach. It actively misteaches. Learners who complete training built on inaccurate content leave with false confidence in false knowledge. In high-stakes domains: healthcare, safety, compliance, financial services, aviation, that false confidence can cause direct harm.

The instructional designer who uses AI-generated content without rigorous accuracy verification is not merely accepting a quality risk. They are accepting a professional responsibility for outcomes they may not be able to foresee. The research citation that doesn't exist. The procedure that sounds authoritative and contradicts the current standard of care. The regulatory guidance that reflects a rule that was updated after the AI's training cutoff. Each of these is a form of intellectual dishonesty, even when it is unintentional, because the designer had the means and the professional responsibility to verify what they were publishing and chose speed over rigor.

The governance response is straightforward to state and demanding to maintain: no factual claim in learning content produced with AI assistance may be published without independent verification from an authoritative source. Not "most factual claims." Not "the important ones." Every factual claim. This standard is not stricter than what responsible content creation has always required. It is the same standard, applied to a context where the temptation to shortcut is much higher because the output looks so much like it should be right.

3. Learner Data Privacy

The AI tools integrated into learning workflows often process learner data, sometimes explicitly and sometimes in ways that practitioners and learners do not fully understand. Survey responses, assessment results, learning behavior patterns, demographic data, and in some systems even written responses to scenario-based activities may be processed by

AI tools whose data handling practices vary significantly and are not always transparent.

The question practitioners must be able to answer about every AI tool in their workflow is not "is it useful?" but "what learner data does it process, where does that data go, how long is it retained, who has access to it, and have the learners who generated it been informed and given meaningful consent?" These are not questions for IT or legal alone. They are questions that instructional designers must understand well enough to make informed decisions about which tools to use, how to use them, and what to disclose to learners.

The regulatory landscape for learner data privacy varies by jurisdiction and sector. GDPR in the European Union, FERPA in US educational institutions, HIPAA where health data intersects with learning content, and COPPA where learners may be minors each impose specific requirements. The designer does not need to be a regulatory expert. They need to know when a data question requires expert consultation and to ask it before deployment rather than after a complaint.

Questions Every Designer Must Be Able to Answer About AI Tools in Their Workflow

- What learner data does this tool process? Does it receive individual names, roles, performance scores, or written responses? Or only aggregate interaction patterns?
- Where is the data stored? In which country or region? Who owns the infrastructure? What are the subprocessor relationships?
- What are the data retention policies? How long is individual learner data held? Is it used to train the tool's underlying models?
- Can learner data be deleted on request? Is there a documented process for data subject access requests?
- What does the organization's privacy notice say about AI processing of learner data? Have learners been informed in terms they can understand?
- Does this tool's use require a Data Protection Impact Assessment under applicable law? Has one been conducted?
- If this tool processes data about minors, does it comply with the age-specific requirements that apply in the relevant jurisdiction?

If you cannot answer these questions for a tool currently in your workflow, that is a governance gap that requires addressing before the next program launches, not after.

4. Intellectual Property and Attribution

AI tools generate content by drawing on patterns learned from vast quantities of existing text, images, and other creative works, much of which was created by people who did not consent to its use in training AI systems and who have not been compensated for the derivative use of their work. The legal status of AI-generated content is actively contested in courts and legislatures as of this writing, and will be for some time. The ethical status is less contested: practitioners who use AI tools that were trained on unlicensed content are participating in a system that has taken value from creators without their consent.

For instructional designers, the IP considerations are practical as well as ethical. When an AI tool generates an image that closely resembles a specific artist's style, or produces text that reproduces distinctive phrasing from a copyrighted source, or creates a scenario that is structurally similar to a published case study, the legal exposure may fall on the organization that deployed the tool, not on the tool's developer. Understanding the indemnification terms of AI tool agreements is therefore not merely a legal department concern. It is a practitioner concern, and practitioners who are not asking these questions of their vendors are accepting risks they may not have been authorized to accept.

5. Consent and Autonomy

The most forward-looking ethical domain in AI-augmented learning is also the one that has received the least attention in current practice: the question of what learners have consented to and what choices they retain in an AI-mediated learning environment.

As AI tools become more deeply integrated into learning delivery, adaptive systems that modify content based on learner behavior, AI tutors that provide individualized feedback, emotion-detection systems that infer engagement from webcam data, predictive systems that flag "at-risk" learners before they have demonstrated any difficulty, the question of what learners know about how they are being observed, what is being inferred from that observation, and what decisions that inference informs becomes genuinely urgent.

Consent in mandatory training contexts is complicated. An employee who is required to complete a compliance program has not, in any meaningful sense, chosen to participate in the learning system that delivers it. The fact that they agreed to employment terms that include training requirements is not the same as informed consent to AI processing of their learning behavior data. The designer who treats the employment relationship as sufficient consent for all downstream data processing is on ethically thin ground, and increasingly on legally thin ground as well.

The principle of learner autonomy, the idea that adult learners have the right to understand the conditions of their learning and to make meaningful choices within them, is not a new principle in instructional design. Knowles articulated it clearly. What is new is the range of ways that AI-mediated learning can impinge on that autonomy without the learner being aware that it is happening. Holding this principle in AI-augmented design requires active attention, not passive assumption.

Building Governance: From Principles to Practice

Ethical commitments without governance structures are aspirations. The gap between the L&D community's stated commitment to inclusive, accurate, trustworthy learning design and the actual practices of teams under production pressure is not primarily a gap in values. It is a gap in systems. Individual practitioners with good intentions, operating without shared standards and review requirements, will produce inconsistent outcomes. Governance structures exist to close that gap.

What follows is a governance framework organized around six principles. Each principle has a practice component, what it means in the day-to-day workflow, and an accountability component, who is responsible for ensuring it is followed. The framework is designed to be adapted, not adopted wholesale: the right governance structure for a solo practitioner differs from the right structure for a twelve-person L&D team, which differs from the right structure for a global learning function. The principles are non-negotiable. The implementation must fit the context.

Principle 1: Human Authorship

AI is a drafting and analysis tool. Every piece of learning content that reaches a learner has been reviewed, evaluated, and approved by a human professional who takes responsibility for its accuracy, fairness, and appropriateness.

In practice: Every AI-generated output that enters the production workflow is tagged at entry. Review documentation specifies what was reviewed, by whom, and against what criteria. The reviewer is named and accountable.

Principle 2: Verified Accuracy

No factual claim in learning content produced with AI assistance is published without independent verification from an authoritative source. This applies to all content regardless of domain, regardless of how confident the AI output appears, and regardless of time pressure.

In practice: Accuracy verification is a named milestone in every project timeline, not an implicit assumption. In specialized domains, verification requires a qualified SME review, not designer judgment alone. Unverified claims are removed or marked as requiring verification before the content moves forward.

Principle 3: Audited Representation

AI-generated scenarios, characters, examples, and images are reviewed for demographic representation patterns before any content is delivered to learners. It examines the full distribution of who appears in what roles across all scenarios in the program, and that should track, whenever possible, with reality.

In practice: A representation audit is a required QA step with the same status as Gagné coverage and objective alignment checks. The audit result is documented.

Principle 4: Disclosed Data Practices

Learners are informed, in language they can understand, about what data is collected during their learning experience, how it is used, and who has access to it. AI processing of learner data is disclosed. Consent is obtained where required and sought where ethically appropriate even when not legally required.

In practice: A data practice disclosure is a standard component of program design, not an afterthought. The learning technologist confirms data handling practices before each program launch. Learners have a mechanism to ask questions and a path to request review of their data.

Principle 5: Proportionate AI Use

AI tools are used where they serve learning quality and efficiency, and not used where their limitations create risks disproportionate to their benefits. The decision to use AI for a specific task in a specific context is a professional judgment call, not a default.

In practice: A "use / don't use" decision framework is documented in the team's AI use policy, covering the content types, domains, and audience characteristics for which AI use requires additional review or is contraindicated. This framework is updated as tool capabilities and limitations evolve.

Principle 6: Living Standards

AI governance standards are reviewed and updated on a defined cadence, because the tools and their limitations change faster than any static policy document can accommodate. The governance framework is a living document owned by a named individual and reviewed at least annually.

In practice: A governance review is scheduled at the start of each year and triggered by significant tool changes or incidents. Review findings are documented and the updated standards are communicated to the full team. New team members receive governance orientation as part of onboarding.

The AI Use Policy: A Template

An AI use policy is the governance framework's operational document: the specific, written agreement between the team and the organization about how AI tools will be used in learning design work. It is distinct from the organization's broader AI use policy, which covers all organizational AI use, in that it addresses the specific applications, risks, and standards relevant to instructional design practice.

The template below is organized as a set of policy sections, each with the content that section should address. It is a starting point, not a finished document: the specific tool names, review processes, and standards must be filled in by the team that owns the policy.

L&D Team AI Use Policy — Template Sections

Purpose	This policy governs the use of artificial intelligence tools in the design, development, and delivery of learning programs by [TEAM/ORGANIZATION]. It establishes standards for responsible AI use that protect learners, ensure content quality, and maintain the professional integrity of our work.
Scope	This policy applies to [NAME ROLES: all instructional designers, developers, facilitators, etc.] and to [NAME TOOL CATEGORIES: generative AI tools, analytics platforms, AI-assisted authoring tools, etc.]. It does not supersede the organization's broader AI use policy; where policies conflict, the more restrictive standard applies.
Approved Tools	The following AI tools are approved for use in L&D work: [LIST TOOLS WITH PERMITTED USE CASES FOR EACH]. Use of AI tools not on this list requires written approval from [NAMED ROLE] before use on any project. The approved tools list is reviewed [QUARTERLY/ANNUALLY] and updated as needed.
Required Review Standards	All AI-generated content used in learning programs must meet the following review standards before being delivered to learners: (1) Accuracy verification against an authoritative source for all factual claims; (2) Representation audit covering demographic patterns across all scenarios and examples; (3) Accessibility review against WCAG 2.1 AA; (4) Domain accuracy review by a qualified SME for specialized content. These reviews are documented in the project record.
Learner Data Handling	AI tools that process learner data [define: assessment responses, behavioral patterns, written submissions, etc.] may only be used where the data handling has been reviewed and approved by [NAMED ROLE / PRIVACY OFFICER]. Learner-facing AI experiences must include a clear disclosure of what data is collected and how it is used. Questions about data handling are escalated to [NAMED ROLE / LEGAL] before program launch.

Disclosure Standards	Learning programs that use AI-generated content [define the threshold: >X% of content, or specific content types] will include a disclosure statement to learners. Programs using AI for adaptive delivery, AI-generated feedback, or real-time AI interaction will include a disclosure statement regardless of content proportion. Disclosure language must be approved by [NAMED ROLE] before publication.

Contraindicated Uses	AI tools will not be used for the following without explicit senior approval: [LIST, e.g., content about mental health, content for vulnerable populations, content with significant legal liability, facial recognition or emotion-detection in delivery]. These restrictions exist because the failure modes in these contexts create risks disproportionate to the efficiency gains.

Incident Response	When an AI-related quality or ethics incident is identified — including bias in deployed content, accuracy errors in published material, unauthorized data processing, or learner complaints about AI-generated content, the incident is reported to [NAMED ROLE] within [TIMEFRAME]. All such incidents are reviewed, documented, and used to inform governance updates. Affected content is reviewed and corrected within [TIMEFRAME].

Governance and Review	This policy is owned by [NAMED ROLE] and reviewed [ANNUALLY / WHEN SIGNIFICANT TOOL OR REGULATORY CHANGES OCCUR]. Team members have the right to raise concerns about AI use practices without retaliation. Questions about this policy are directed to [NAMED ROLE].

A policy document only governs behavior if it is known, understood, and held to. Building the policy is the beginning. Orienting new team members to it, referencing it when review standards are under pressure, and updating it when the tools or the risks change are the ongoing practices that make it real.

The Practitioner's Posture

Governance frameworks address organizational practice. This section addresses individual posture: the orientation that individual practitioners bring to their AI-augmented work, regardless of whether their organization has a formal governance framework. The practitioner's ethical posture in AI-augmented design has three components that no policy document can supply.

Intellectual Humility About What AI Produces

The output of an AI tool is a starting point, not a conclusion. It may be accurate. It may be biased. It may be plausible without being true. It may be fair on its surface and systematically unfair in its patterns. The practitioner who approaches AI output with the assumption that it requires critical evaluation, not as a formality, but as a genuine intellectual act, is practicing intellectual humility. The practitioner who approaches it with the assumption that it is probably fine is practicing the abdication of professional responsibility.

This humility is not about distrust of the tools. It is about accurate understanding of what the tools are: statistical systems that produce likely outputs, not knowledge systems that produce true ones. The outputs are often very good. They are never inherently trustworthy. Maintaining that distinction in practice, under time pressure, when the output looks exactly like what you needed and the deadline is tomorrow, is what intellectual humility actually requires.

Advocacy for the Learner

Every practitioner in this field has a primary client relationship with a stakeholder, a department head, an HR director, a CHRO, a client organization. The financial relationship is with that stakeholder. The professional responsibility runs further: It runs to the learner.

The learner is the person whose competence, confidence, and professional development are affected by the quality and integrity of what the practitioner builds. The learner is the person who will encounter the bias in the scenario generation if the representation audit is skipped.

The learner is the person who will carry false confidence in false information if the accuracy review is abbreviated. The learner is the person whose data will be processed in ways they were not informed about if the disclosure is omitted.

Learner advocacy in AI-augmented practice means asking, about every significant design decision: who is this for, and does this serve them well? Not in the abstract. Specifically: does this scenario represent the learners who will encounter it fairly? Does this assessment measure what they actually need to be able to do? Does this data collection serve their development or just the organization's reporting needs? Is this AI-generated content I would be comfortable with if my name were on it and a learner complained?

The Willingness to Slow Down

Every ethical failure in AI-augmented instructional design has a precondition: someone was moving too fast to notice what was wrong, or moving too fast to care, or moving too fast because the organizational pressure made slowing down feel impossible.

The customer service bias scenario at the opening of this chapter was not caught because no one reviewed the demographic patterns in the scenarios before the program launched. That review would have taken perhaps two hours. The post-launch revision took two weeks, involved a formal apology to the new hire who flagged it, and left a record in the organization's memory that will outlast the program itself.

The willingness to slow down is not always available to practitioners under genuine deadline pressure. It is always a professional responsibility. When slowing down is not possible within the available time and resources, the appropriate response is not to proceed anyway. It is to name the constraint explicitly, document the risk, and give the decision about whether to proceed to the person who has the authority to accept it. That person may decide to proceed. That is their right. The practitioner's right, and responsibility, is to ensure that the decision is made with full knowledge of what is being risked.

When You See Something

The chapter closes with a practical question that the governance frameworks and ethical principles above cannot fully answer: what do you do when you see something wrong and the organizational context makes it difficult to name it?

This is not a hypothetical scenario. Practitioners regularly encounter situations where a program has a bias problem that leadership does not want to hear about, where an accuracy issue has been flagged and dismissed, where a data practice is questionable and the legal team has approved it despite the ethical concern, where a stakeholder has requested content that the designer believes is harmful and the organizational incentives point toward compliance rather than resistance. These situations require the same courage that Chapter 13 called the performance consultant's essential competency: the willingness to say what you see, clearly, with evidence, to the person who needs to hear it.

A few principles for these moments, drawn from practitioners who have navigated them:

- **Name the specific harm, not the general principle.** "I'm concerned about the ethical implications of this approach" is easy to dismiss. "The demographic pattern in these scenarios will teach a specific association to eight hundred employees, and here is what that association is" is harder to dismiss, because it is specific enough to evaluate. Ethical concerns land when they are concrete.
- **Document before you speak.** If you have identified a problem and are about to name it to a stakeholder who may not receive it well, document what you found, when you found it, and what you recommended before the conversation happens. This documentation is not primarily for legal protection. It is for professional integrity: a record that you did your job.
- **Know your escalation path.** In most organizations, the instructional designer is not the final decision-maker on whether a program launches. When a concern is dismissed at one level, the appropriate next step is not silence. It is escalation to the next level of decision authority, with the same specificity and documentation. Know who that is before you need to know it.

- **Be willing to be the person who slowed it down.** The practitioner who flags a bias problem and causes a two-week delay will sometimes not be thanked for it. They will be thanked by the eight hundred employees who encountered the corrected content instead of the original. That accounting is not always visible at the time of the decision. It is real, nonetheless.

The New Hire's Feedback Form

The organization that received the new hire's feedback form did two things. The first thing they did was fix the program: a complete scenario revision, a representation audit applied to all eight modules, and a new QA checkpoint added to the L&D team's production process. The revision took three weeks.

The second thing they did was less visible but more important. They sent a personal response to the new hire who had submitted the form. Not a form letter acknowledging receipt. A specific response from the L&D director that named what the new hire had seen, confirmed that it had been a real problem and not a misreading, explained what was being done about it, and thanked her for saying something.

The new hire later said, in a team meeting at the end of her first year, that the response had been one of the things that convinced her the organization was worth staying at. Not because the program had been wrong, programs made mistakes, she understood that, but because the response to the mistake had been honest and specific and treated her as someone whose observation mattered.

This is what ethical practice in AI-augmented learning design looks like when it is working: not the absence of error, which is not a realistic standard for any human endeavor. The presence of the systems, the habits, and the culture that catch errors before they compound, that respond to the ones that get through with honesty rather than defensiveness, and that treat the learner's experience of the content, including the experience of seeing something wrong and naming it, as information worth acting on.

The practitioner's responsibility in this chain is not to be perfect. It is to be the person who takes that responsibility seriously, in the work, and in the moment when it would be easier not to.

Chapter Summary

- The ethical stakes of AI in learning are active, not hypothetical. Learning content shapes how people think; when that content is produced with AI tools that encode bias, generate inaccuracies, or handle data without proper disclosure, the harm is measurable in learner experience and organizational practice.
- The five ethical domains require active governance: (1) Bias, demographic defaults and uneven attribution; (2) Accuracy and intellectual integrity, the professional responsibility to verify; (3) Learner data privacy, what is collected, where it goes, who consented; (4) Intellectual property and attribution, disclosure and legal exposure; (5) Consent and autonomy, what learners know about how they are being observed.
- The six-principle governance framework, human authorship, verified accuracy, audited representation, disclosed data practices, proportionate AI use, and living standards, translates ethical commitments into operational practices with named accountability.
- The AI use policy template provides the organizational document that makes governance enforceable: specific tools, required review standards, data handling protocols, disclosure requirements, contraindicated uses, and incident response. It is a starting point that must be adapted to the team's context.
- The practitioner's posture has three components that no policy can supply: intellectual humility about what AI produces, advocacy for the learner above the organizational client, and the willingness to slow down when something requires more time than the schedule allows.
- When you see something wrong, the responsible path is specific documentation, specific communication, and a known escalation path, not silence. The designer who builds something they know has a problem and says nothing has made a decision about whose interests matter more.

Reflection Questions

For individual journaling or team discussion.

1. The customer service bias scenario in this chapter's opening was not caught by multiple rounds of review. Apply the representation audit to a recent program you developed: does the demographic distribution across all your scenarios hold up to examination? What pattern emerges when you look across the full program rather than at individual screens?
2. Work through the data privacy questions in this chapter for each AI tool currently in your workflow. How many can you answer? For the ones you cannot, what would it take to get answers, and who would you need to involve?

3. Does your team or organization have an AI use policy? If yes, when was it last reviewed, and does it address the five ethical domains in this chapter? If no, what is the smallest viable version of a policy you could draft and propose in the next thirty days?
4. Think of a time when you saw a quality or ethics problem in AI-assisted work, your own or a colleague's, and did not name it, or named it and it was dismissed. What prevented the problem from being addressed? What would have needed to be different, in your approach, in the organizational context, or in both, for the outcome to be different?

Part Four · Chapter 18

Leading the Transition: From ADDIE to INDIE in Your Organization

The all-team meeting had gone well, which made what came after harder. Silas, the L&D director, eighteen months into his role at a mid-sized financial services organization, had presented the INDIE framework to his team of nine the previous Tuesday. He had read widely, prepared carefully, and delivered what he believed was a genuinely compelling case for why the team needed to evolve its practice. The slides were good. The rationale was sound. Several team members had asked engaged questions. One had stayed afterward to say she was excited.

Six weeks later, nothing had changed.

Not visibly, anyway. Two of the nine team members were using AI tools more than they had before, which was good. The projects they were using them on were still organized around ADDIE phases and ADDIE deliverables, because that was what the project templates required and nobody had updated the templates. The "excited" team member had attempted the Phase-IN fusion on a small project, encountered a stakeholder who expected a formal analysis document before any design work began, and reverted to the familiar process without telling Silas. The team member she had mentioned her difficulty to had told her that this was what always happened when someone tried to change things. She had filed that observation away and moved on.

Silas sat with this for a while before he understood what he had done. He had made a presentation about change. He had not made a change.

Why Transitions Fail: The Six Patterns

The gap between an organization's stated commitment to a new framework and its actual adoption of that framework is one of the most familiar and most frustrating experiences in organizational life. It is also one of the most predictable. The failure modes of change initiatives are not random. They cluster into a small number of patterns that repeat across industries, organizational sizes, and change types with enough consistency to be anticipated, named, and addressed before they occur.

What follows is an honest account of the six most common reasons INDIE adoption stalls, based on the patterns observed in teams that have attempted structured practice transitions. The chapter that addresses them is this one. The leader who can name them in advance is the leader who will not be surprised when they arrive.

The Presentation That Substitutes for Change

What it looks like: A compelling kickoff session generates genuine enthusiasm. Six weeks later, nothing is different. People remember the presentation fondly. They have not changed what they do.

Why it happens: Presentations create understanding and sometimes motivation. They do not change habits, templates, workflows, or the organizational systems that reinforce existing behavior. Understanding INDIE is necessary but not sufficient for practicing it.

What actually helps: Replace the kickoff presentation with a kickoff experiment: instead of telling the team what INDIE is, run a live demonstration of the Phase-IN fusion on a real project question during the meeting. Let people experience the difference, nor just hear about it.

The Infrastructure Gap

What it looks like: Practitioners are enthusiastic about INDIE in principle and frustrated in practice because the project templates, approval processes, and stakeholder expectations are all built around ADDIE. The new framework doesn't fit the existing containers.

Why it happens: Organizational practice is sustained by organizational infrastructure. Timelines that assume sequential phases, templates that require an analysis deliverable before design can begin, and stakeholder expectations built around ADDIE review points all create structural resistance to INDIE adoption that individual motivation cannot overcome.

What actually helps: Update the infrastructure before announcing the framework. Revised project templates, updated stakeholder communication guides, and adjusted timeline standards should be ready before INDIE is introduced. Practitioners should not have to fight their own tools to practice the framework they've been asked to adopt.

The Uneven Adoption Spiral

What it looks like: A few enthusiastic early adopters make rapid progress. The rest of the team watches, some with interest and some with skepticism. The gap between early adopters and the rest widens until it becomes a visible capability divide that creates resentment on both sides.

Why it happens: Adoption initiatives almost always produce early adopters. Without a deliberate structure for sharing what early adopters are learning, the initiative fragments into a two-speed team in which the early adopters pull ahead and the rest fall behind. The early adopters may not notice. The rest will.

What actually helps: Design peer learning into the adoption from week one. A structured monthly session in which early adopters share one specific thing they learned is not optional community-building. It is the mechanism that prevents the adoption from becoming a capability divide.

The Stakeholder Ambush

What it looks like: A practitioner attempts an INDIE-informed approach on a project. A stakeholder who was not informed of the transition pushes back hard, citing the expected deliverable structure that ADDIE provided. The practitioner reverts to ADDIE. The story spreads through the team.

Why it happens: Stakeholders whose working relationship with the L&D team is built around ADDIE's deliverable structure will experience INDIE as L&D doing something differently without telling them. That is exactly what it is, unless the stakeholder communication has been done deliberately and in advance.

What actually helps: Stakeholder communication is not a courtesy. It is a precondition for successful adoption. Every stakeholder who regularly reviews L&D work needs to know, before any INDIE-informed project begins: what will change, what will stay the same, and what they can expect instead of the familiar deliverables.

The Quality Backslide

What it looks like: Early AI-assisted work is faster but noticeably uneven. Some outputs are excellent. Others reveal that the review processes have not kept pace with the production speed. A stakeholder notices. The team's credibility takes a visible hit.

Why it happens: AI adoption without quality governance is velocity without direction. When production speed increases without corresponding attention to review standards, the outputs that reach stakeholders will occasionally be worse than what the team produced before AI, because the old slow process, whatever its other failings, built in enough time for the designer to catch their own errors.

What actually helps: Establish the AI quality review standard before production volume increases. The Development QA checklist from Chapter 7 and the governance principles from Chapter 17 are not bureaucratic overhead. They are what keeps increased speed from producing decreased quality.

The Momentum Cliff

What it looks like: The first ninety days go well. The team makes genuine progress. Then the next urgent project arrives, and the one after that, and the structured reflection sessions get deprioritized, the prompt library stops being updated, and twelve months later the team is operating at roughly the same level of AI integration as they were at month three.

Why it happens: Organizational change without a sustaining rhythm collapses under operational pressure. The activities that sustain adoption, peer learning sessions, prompt library maintenance, governance review, team reflection, are all low-urgency and high-importance. They will always lose to high-urgency and high-importance when a leader doesn't protect them.

What actually helps: Build the sustaining rhythm into the organizational calendar before it can be crowded out. The monthly community of practice session, the quarterly prompt library review, and the annual governance assessment should be standing calendar items with the same protected status as client delivery milestones.

The Conditions for Successful Transition

Understanding the failure patterns is necessary but not sufficient. What follows are the conditions that distinguish transitions that actually change how teams work from transitions that are remembered as a good initiative that didn't quite land. These conditions are not sequential steps. They are concurrent requirements: a transition that meets some of them but not others will encounter the failure patterns it didn't address.

Leadership Behavior, Not Leadership Announcement

The most reliable predictor of whether a framework transition succeeds is not the quality of the kickoff presentation. It is whether the leader who announced the transition visibly practices it in their own work. A leader who introduces INDIE and then continues to manage projects through ADDIE-structured templates, continues to request analysis deliverables before design work begins, and continues to evaluate L&D quality against ADDIE-informed standards has told the team, through their behavior, that the announcement was aspirational rather than operational.

Modeling is not a leadership technique. It is the primary mechanism through which organizational culture actually changes rather than merely announces itself. The L&D leader who wants their team to practice INDIE must themselves practice INDIE, in how they conduct project conversations, in how they use AI tools in their own work, in how they structure the team's shared work, and in how they talk about quality with stakeholders. The team will watch what the leader does, not what the leader said at the kickoff.

Infrastructure Before Announcement

The infrastructure gap failure pattern is almost entirely preventable through sequencing. A team cannot practice a new framework using containers designed for the old one. Project templates, timeline standards, stakeholder communication guides, quality review checklists, and prompt libraries should all be adapted to the new framework before practitioners are asked to use it. This is not a small amount of work. It is the work that makes adoption possible rather than aspirational.

The practical sequence: the leader and one or two trusted practitioners develop the infrastructure over three to four weeks before the team-wide adoption announcement. The infrastructure is tested on at least one live project before it is presented to the full team as the new standard. The announcement includes the new templates and tools, not just the new ideas.

Stakeholder Preparation

Every stakeholder who regularly reviews, approves, or receives L&D work needs a specific, tailored communication about what the transition means for their working relationship with the team. Not a broadcast announcement. A conversation, or at minimum a personal message that explains what they can expect to see differently, what will remain the same, and what the change is designed to produce for them.

The stakeholder communication has two purposes. The obvious one is managing expectations: preventing the ambush scenario in which a stakeholder encounters a new deliverable structure without warning and interprets it as the team doing something irregular. The less obvious purpose is equally important: it creates an opportunity to name what the transition is designed to improve, which positions the stakeholder as a beneficiary of the change rather than a recipient of it. Stakeholders who understand that INDIE's iteration model will produce better programs faster are stakeholders who may actively support the transition rather than merely tolerating it.

PROMPT: Stakeholder Transition Briefing
I need to prepare personalized communications for stakeholders as my L&D team transitions to the INDIE framework. For each stakeholder below, generate a tailored briefing message: Stakeholder: [NAME / ROLE] Current working relationship with L&D: [DESCRIBE — what they request, review, and receive from the team] What will change for them: [DESCRIBE SPECIFICALLY] What will not change: [DESCRIBE] The benefit they can expect: [DESCRIBE IN THEIR TERMS, NOT L&D TERMS] Any sensitivity in this relationship: [DESCRIBE]

> Please draft a communication that:
> 1. Opens with what matters to them (the business outcome), not with what matters to the L&D team (the process change)
> 2. Names specifically what they will see differently
> 3. Explicitly names what they will continue to receive that they value
> 4. Invites a conversation rather than announcing a decision
> 5. Is calibrated to the relationship (more formal for senior executives, more collegial for regular working partners)
>
> Generate one message per stakeholder. Do not use a template that reads like a template. Each message should feel specific to that relationship.

Permission to Be Imperfect

Teams that are afraid of doing INDIE wrong do not do INDIE. They do ADDIE with INDIE-sounding language in their project documents, which is a worse outcome than simply continuing to practice ADDIE. The leader's job is to create explicit permission for the learning period that precedes competence: the period when the team is trying the framework on real projects, making mistakes, and learning from them before they have developed the judgment to apply the framework fluently.

This permission must be named explicitly, because the implicit organizational message in most L&D teams is the opposite: that quality is the standard and anything short of quality is a failure. That message is appropriate for work going to clients. It is counterproductive for an internal learning process about a new framework. The leader who creates explicit protection for the learning period, who names in advance that the first three months of INDIE adoption will include mistakes, that those mistakes are expected and valued as information, and that they will not be treated as performance failures, is the leader whose team will actually attempt the transition.

Structured Learning, Not Just Encouraged Learning

The monthly community of practice session is the single most important sustaining structure for INDIE adoption. Not because it provides formal training, it doesn't, but because it creates the regular, structured opportunity for practitioners to share what they are learning with each other rather than accumulating that learning in individual silos.

The session format matters. A status update meeting is not a community of practice session. A workshop on a specific AI tool is not a community of practice session. A thirty-minute meeting in which each practitioner shares one thing that worked, one thing that didn't, and one question they are currently sitting with, that is a community of practice session. It is structured enough to produce consistent output and open enough to surface the unexpected learning that structured agendas suppress.

The Full Roadmap: A Complete Transition Plan

The 90-day roadmap in Chapter 14 addressed INDIE adoption for an L&D team. This section extends that roadmap to the full organizational transition: from the leader's own preparation through the team adoption through the stakeholder integration through the sustained rhythm that makes the adoption durable. It is organized as a week-by-week guide rather than a phase description, because the specificity of a week-by-week plan is what makes the difference between a roadmap that is followed and one that is filed.

Weeks 1–2	**Leader Preparation** ▶ **Actions:** Read this book, including the role chapters relevant to your team composition. Identify the three failure patterns most likely to affect your specific team and organizational context. Map the infrastructure that needs to be updated before adoption begins. Identify your one or two most trusted practitioners to involve in the infrastructure development. ▶ **Outputs:** Personal INDIE framework understanding. Honest assessment of your team's current AI maturity level. Preliminary infrastructure gap analysis.

Weeks 3–4	**Infrastructure Development** ▶ **Actions:** With your trusted practitioners, develop: updated project templates (IN phase combined, Enhancement cycle built in), revised timeline standards, stakeholder communication templates for the transition, first version of the shared prompt library (starting with the prompts from this book's appendix), and the AI quality review standard your team will use. ▶ **Outputs:** Updated project templates. Revised timeline standards. Transition communication templates. Version 1.0 of the shared prompt library. Draft AI quality review standard.

Weeks 5–6	**Pilot Project and Stakeholder Preparation** ▶ **Actions:** Run one real project through the new templates and process. Document what works and what requires adjustment. Simultaneously, conduct stakeholder briefing conversations with your three to five most important stakeholder relationships. Do not brief all stakeholders at once, start with the ones whose working relationship with L&D is most critical and most likely to shape others' perceptions. ▶ **Outputs:** One project running under INDIE framework. Documented learnings from pilot. Briefing conversations completed with priority stakeholders. Stakeholder feedback incorporated into framework adaptation.

Weeks 7–9	**Team Introduction** ▶ **Actions:** Introduce the framework to the full team — not as a presentation, but as a structured working session using the pilot project as the teaching example. Show the team what changed, what was learned, and what the new infrastructure looks like. Name explicitly: this is a learning period, mistakes are expected, the goal is to have everyone operating comfortably in the new framework by the end of month six. Assign each team member a specific INDIE phase chapter to own and present to the team in the following month. ▶ **Outputs:** Full team introduction to INDIE using real project evidence. Team members assigned chapter ownership. Learning period expectations named explicitly. Schedule set for monthly community of practice sessions.

Weeks 10–12	**Initial Adoption and First CoP Sessions** ▶ **Actions:** All active projects begin operating under the new framework. Run the first two monthly community of practice sessions — one practitioner presents their chapter ownership, then the full group shares learnings from the first month of adoption. Update the prompt library with contributions from the first month. Identify the practitioners who are struggling and provide specific support rather than general encouragement. ▶ **Outputs:** All projects on INDIE framework. Prompt library updated with first contributions. Two CoP sessions completed and documented. Specific support provided to struggling practitioners.

Months 4–6	**Consolidation and Standard Development** ▶ **Actions:** Continue the monthly CoP sessions. By month six, the framework should feel like the team's way of working rather than an initiative being followed. Conduct a team self-assessment against the AI maturity framework from Chapter 14. Update the AI use policy based on three months of experience. Identify the practices that are working best and document them for onboarding new team members. ▶ **Outputs:** AI maturity assessment completed. AI use policy updated. Onboarding documentation for new team members. Month-six retrospective: what has genuinely changed, what hasn't, and why.

Month 12	**Annual Review and Sustained Rhythm** ▶ **Actions:** Conduct the first annual INDIE and AI governance review: framework effectiveness against portfolio quality data, AI maturity level compared to month one, prompt library health, policy currency against tool and regulatory changes, and the honest answer to: is the team actually doing INDIE, or doing ADDIE with INDIE-sounding language? Update the framework, the tools, and the standards as needed. Celebrate the practitioners who led the way. ▶ **Outputs:** Annual review document. Updated framework standards. Updated AI governance policy. Recognition of practitioners who drove the adoption. Roadmap for year two.

Making the Organizational Case

For many L&D leaders, the transition to INDIE requires not just internal team adoption but organizational investment: budget for AI tools, protected time for the community of practice sessions, approval to update the project deliverable structure that stakeholders expect, and sometimes headcount or skill development investment to close capability gaps the transition reveals. Making that case effectively requires the same evidence-based approach that the Enhancement phase applies to program quality.

The organizational case for INDIE adoption does not lead with the framework. It leads with the problem the framework solves. Executives who approve investments in L&D process change are not persuaded by arguments about instructional design theory. They are persuaded by arguments about organizational outcomes: programs delivered faster, evidence of learning effectiveness that currently doesn't exist, reduced rework cost from late-stage quality failures, better alignment between learning investment and business priorities. INDIE's case for adoption is built from exactly these outcomes, and the leader who makes the case in these terms is the leader who is likely to receive the investment.

PROMPT: Organizational Case for INDIE Adoption
I need to make a case to [CHRO / CISO / EXECUTIVE SPONSOR] for investing in the transition to the INDIE framework and AI-augmented L&D practice. Current state: Team size and composition: [DESCRIBE] Current process: [DESCRIBE ADDIE-BASED OR OTHER CURRENT APPROACH] Known quality or efficiency problems: [DESCRIBE HONESTLY]

```
 Current AI tool usage: [DESCRIBE]
 Portfolio size: [NUMBER OF ACTIVE PROGRAMS]

Investment required:
 Tool costs: [DESCRIBE]
 Time investment: [HOURS FOR TRANSITION, ONGOING CoP SESSIONS, ETC.]
 Any headcount or capability development: [DESCRIBE]

Available evidence:
 Current program quality metrics: [DESCRIBE WHAT EXISTS]
 Current delivery timelines: [DESCRIBE]
 Known enhancement gaps: [PROGRAMS WITHOUT EVALUATION DATA, ETC.]

Please build a business case that:
 1. Opens with the organizational cost of the current state
   (slow delivery, quality inconsistency, programs without evidence,
   team capability gaps) in terms the audience cares about
 2. Describes INDIE as a solution to those specific problems, not
   as an upgrade to L&D process for its own sake
 3. Connects the investment to specific, measurable outcomes
   with a realistic timeline for seeing them
 4. Addresses the risks of the transition honestly
   (the failure patterns in this chapter) and how they will be managed
 5. Makes a specific request: what decision or approval is needed,
   by when, to make this happen

Avoid L&D jargon. Assume an audience that values evidence, directness,
and honest acknowledgment of risk over polished optimism.
```

What an INDIE-Native Team Looks Like

The roadmap describes the transition. This section describes the destination: what a team looks and operates like when INDIE has been genuinely adopted rather than nominally adopted. It is not a perfection portrait. It is an honest description of the practices, habits, and culture that distinguish a team that has internalized the framework from one that is still learning it.

How long does it take to get there? The honest answer is twelve to eighteen months for a team that is actively working at it. The month-six assessment should show meaningful progress. The month-twelve assessment should show genuine fluency. Full organizational integration, where the framework is how the team works rather than what the team is trying to do, typically requires the second year.

Project initiation	The IN phase begins with a brief integration conversation that includes the learning technologist and identifies any technical constraints before design work begins. Analysis and design happen concurrently from the first client conversation, not in sequence.
AI use	AI tools are used for specific, defined tasks in every project. AI output is reviewed against written standards, not just individual judgment. The shared prompt library is consulted before any new prompting task and updated when better approaches are found.
Stakeholder management	Stakeholders receive pilots rather than final programs. Pilot feedback is synthesized within a week and acted on. Stakeholders understand the Enhancement cycle and expect data-informed iteration rather than program completion.
Quality culture	Quality review of AI output is a workflow step, not a personal choice. Representation audits happen on every program before delivery. The Development QA checklist is used systematically, not selectively.
Enhancement practice	Every active program has an Enhancement dashboard. Level 1 data is synthesized within 24 hours of each delivery. Level 3 follow-up is designed in the IN phase, not after launch. The team can answer "what is this program achieving?" for each active program at any time.
Team learning	The monthly CoP session happens without being rescheduled for project pressure. Practitioners share what is and isn't working with the expectation that the team will learn from it collectively. The prompt library is updated monthly.
AI governance	The AI use policy is current and known. Representation audits are documented. Data privacy questions are routed to the appropriate reviewer before any program uses a new AI tool. The governance review happens annually.

Leadership credibility	The L&D leader can answer "what are we getting for what we're spending?" with specific data for the portfolio's most important programs. The retirement/invest/maintain portfolio decision is made on evidence, not tenure. The team is known in the organization for producing programs that work, not just programs that get delivered.

Silas, Six Months Later

Silas spent the month after his realization doing the work he had skipped. He updated the project templates himself, working with two team members who had already been experimenting with the Phase-IN fusion. He had the stakeholder conversations, one of which revealed that his most important business unit partner had actually been frustrated with the sequential ADDIE process for years and had been waiting for someone to propose something different. He ran a small project through the new templates and documented what worked and what needed adjustment. He sent the new project guide to the full team with a note that said, specifically: "This is the infrastructure I should have built before I introduced the framework. I got that order wrong. Here it is now."

The team meeting when he introduced the framework for the second time was different from the first one. He did not present slides. He walked the team through the actual project he had run on the new templates, showed them what the concurrent analysis-design streams had looked like, and had the team member who had attempted the Phase-IN fusion and run into the stakeholder problem describe what had happened and what she would do differently now. She did this with a directness that was visibly uncomfortable and that the team received with the attention of people recognizing something true.

Six months after the second meeting, the team's AI maturity assessment showed meaningful progress across the board. The prompt library had forty-seven entries. Three programs were running Enhancement cycles. Silas's quarterly review with the CHRO used data he had not previously been able to produce. The "excited" team member had become the team's most consistent practitioner of the Phase-IN fusion and had started mentoring a newer team member in the approach.

None of this had happened because of the first presentation. It had happened because of the infrastructure, the stakeholder preparation, the permission to be imperfect, and the monthly session that kept the learning moving in a direction. It had happened because Silas had understood what he was actually asking for and had done the work to make it possible.

That is what leading a transition actually requires. Not a better presentation. The work before the presentation, and the sustained attention after it, that turns a good idea into a different practice.

Chapter Summary

- The six failure patterns of INDIE adoption, the presentation that substitutes for change, the infrastructure gap, the uneven adoption spiral, the stakeholder ambush, the quality backslide, and the momentum cliff, are predictable and addressable if named in advance. Each has a specific response that prevents or recovers it.
- The conditions for successful transition are concurrent requirements, not sequential steps: leadership behavior (not announcement), infrastructure before announcement, stakeholder preparation, explicit permission to be imperfect during the learning period, and structured peer learning rather than merely encouraged learning.
- The full transition roadmap runs from weeks one through twelve and into the first annual review: leader preparation, infrastructure development, pilot project and stakeholder preparation, team introduction, initial adoption, consolidation, and sustained rhythm. The specificity of week-by-week planning is what makes the difference between a roadmap that is followed and one that is filed.
- The organizational case for INDIE adoption leads with the problems it solves (slow delivery, quality inconsistency, programs without evidence), not with the framework itself. Executives approve investments in outcomes, not in process improvement theory.
- An INDIE-native team is recognizable by its practices, not its vocabulary: AI use governed by shared standards, Enhancement cycles running on active programs, monthly CoP sessions that actually happen, stakeholders receiving pilots instead of launches, and a leader who can answer "what are we getting for what we're spending?" with specific data.
- The transition takes twelve to eighteen months to produce genuine fluency. Month-six assessment should show meaningful progress. Month-twelve assessment should show genuine fluency. Full organizational integration requires the second year. Leaders who expect transformation in ninety days will produce the momentum cliff failure pattern.

Reflection Questions

For individual journaling or team discussion.

1. Silas's fundamental error was making a presentation about change rather than making a change. Have you made this mistake, or observed a leader making it? What specifically distinguished the announcement from the actual work that was required? What would have needed to be different for the change to take root?
1. Of the six failure patterns, which one is most likely to affect a transition you are leading or planning to lead? Name specifically why you think that pattern will emerge, and what you would do in the three weeks before the transition announcement to address it.
2. Apply the infrastructure audit to your current organizational context: which project templates, timeline standards, and stakeholder expectations are built around ADDIE and would need to change before INDIE adoption could succeed? What is the realistic effort required to update them?
3. Think about the stakeholder relationships that most need preparation before an INDIE transition in your organization. For the two most important ones, what do they currently expect from L&D, what would change under INDIE, and how would you frame the change as a benefit rather than an imposition?

Part Five · Conclusion

The Designer in the Age of Intelligent Systems

There is a question that runs beneath every chapter in this book, usually unstated, occasionally surfacing directly: what is instructional design actually for?

Not what does it produce. Not what process does it follow. What is it for, at the level of purpose, the reason it exists as a discipline, the human need it serves, the thing it would leave undone if it disappeared?

The answer the field has historically given is some version of this: instructional design exists to help people learn things that matter, in ways that actually work, so that they can do things they could not do before, or do things they already did with greater competence, confidence, and care. That is the work. Everything else, the frameworks, the theories, the tools, the workflows, is in service of that purpose, not the other way around.

AI has not changed this purpose. It has changed the conditions in which the purpose is pursued. And in changing those conditions, it has made the purpose both more achievable and more demanding of the specifically human capacities that AI does not have: judgment, ethics, relationship, and the stubborn commitment to asking whether the thing being built is actually serving the person who will encounter it.

What Is Coming

This book has addressed the current generation of AI tools: language models that draft, synthesize, analyze, and generate at a speed and scale that has already materially changed instructional design practice. What is coming is more consequential, and practitioners who understand the current generation are better positioned to engage with the next, not because the specific tools will be the same, but because the questions they raise will be.

Adaptive Learning Systems

The next significant AI application in learning is not a better drafting tool. It is the adaptive learning system: a delivery environment that adjusts content, pacing, difficulty, and pathway in real time based on continuous inference about the individual learner's current state. Adaptive systems exist today in limited forms. What is coming is more capable, more granular, and more deeply integrated into how learning experiences are delivered.

Adaptive systems change the designer's role in a specific way: they shift the designer's focus from constructing a fixed learning sequence to designing a learning space with boundaries and affordances. Instead of "the learner will encounter Event 1, then Event 2, in this order," the designer asks: "What must this system always include? What can it vary? What are the learning goals that constrain the variation? What does a learner need to have experienced before they encounter this content, regardless of which path brought them here?" This is harder design work than fixed-sequence design, not easier. It requires deeper knowledge of the learner population, more precise articulation of objectives, more sophisticated assessment architecture, and more careful thinking about what a good learning experience means across the full range of possible paths. AI will handle the real-time adjustment. The designer will define the space it adjusts within.

AI Tutors and Personalized Learning at Scale

AI tutoring systems, applications that provide individualized guidance, feedback, and explanation in response to learner input, in real time, at a quality approaching what a skilled human tutor provides, are closer to practical deployment than most practitioners currently appreciate. The arrival of AI tutors does not eliminate the instructional designer's role. It concentrates it. The designer who has been responsible for building a fixed learning experience becomes responsible for defining the pedagogical logic that the AI tutor enacts: the learning goals it pursues, the misconceptions it is designed to address, the boundaries of its subject matter expertise, the escalation criteria that determine when the learner needs a human instructor rather than an AI one, and the ethical guardrails that govern how the system responds to vulnerable or struggling learners. These are design decisions. They require instructional design expertise. They will determine whether AI tutoring systems produce the kind of learning they could produce, or the kind that is technically functional and educationally mediocre.

Real-Time Performance Support

The most transformative near-term AI application for organizational learning may not be in training at all. It may be in performance support: AI systems embedded in the work environment that provide guidance, information, and decision support at the moment of need, without requiring the worker to leave the work and go to training. Job aids, performance support tools, and embedded help systems have existed for decades. What AI changes is the quality and specificity of the support, a system that can understand a particular situation described in natural language, retrieve the relevant guidance, synthesize it into a useful answer, and deliver it in seconds.

The performance consultant's question from Chapter 13 becomes, in this context, even more pressing: if real-time AI support can provide the information a worker needs at the moment they need it, what does training do that performance support cannot? The honest answer is that training develops the judgment, conceptual understanding, and procedural fluency that allows a worker to recognize what kind of situation they are in and what kind of support they need, including whether they need support at all, or whether they already know. AI performance support serves the competent practitioner. Training builds the competence. Both will be necessary, and the designer who understands the relationship between them will design both more effectively than the designer who treats them as alternatives.

> *As learning systems become more intelligent, the designer's role becomes more important, not less. Someone has to define what that intelligence should be in service of.*

The Argument About What Doesn't Change

Every technology wave in the history of instructional design has produced a version of the same argument: this technology will replace the instructional designer. Film will replace teachers. Television will replace classrooms. eLearning will replace instructor-led training. Adaptive systems will replace curriculum designers. AI tutors will replace human tutors. Each time, the argument has been partially right about what the technology can do and consistently wrong about what that means for the humans who design learning.

The pattern is not that technology fails to do what it promises. Often it succeeds. The pattern is that doing what it promises reveals new requirements for the human expertise that the technology depends on to do it well. Film required cinematographers, directors, and editors whose craft had not previously existed. eLearning required designers who understood interaction design, accessibility, and the particular ways digital delivery changes learner engagement. Adaptive systems require designers who can specify the pedagogical logic of a dynamic learning space in ways that none of the previous generation of design work required.

The technology does not replace the expertise. It changes what the expertise is applied to. And the expertise that survives every technology wave is always the same expertise: the ability to understand what a learner needs, to design experiences that meet that need, to evaluate whether those experiences are working, and to hold the standard of effectiveness against the pressures that push toward efficiency, volume, and the appearance of learning rather than its substance.

That expertise is not threatened by AI. It is made more necessary by it. As learning systems become more capable of delivering content, adapting pathways, and providing feedback without human involvement, the question of what those systems should be doing, what goals they are pursuing, whose interests they serve, what they teach by implication when they are not teaching by intention, becomes more important. Those are not technical questions. They are design questions. They are, at their deepest, questions about human development and human dignity that require human judgment to answer.

The Practitioners in This Book

The practitioners whose stories run through these chapters are composites. No single person experienced all of what is attributed to any one of them. But the experiences they represent are real, drawn from the accumulated observation of instructional designers, facilitators, performance consultants, L&D managers, and organizational development specialists who are navigating this transition in organizations of every size and sector.

Maria, who sat at her desk with a cup of cold coffee and a learning objective that took thirty seconds, and a question she didn't yet know how to answer. James, who needed a plain account of something that kept being explained in the wrong language. Diana, who was quietly terrified and wouldn't say so professionally. Sofia, who finally understood that the prompt was not the variable. Marcus, who did the arithmetic in the conference room and said what needed saying. Ruth, who walked into the meeting not with a blank intake form but with a working hypothesis. Rafael, who sent the storyboard back with a single comment: "This is organized. It is not instructional." Sonia, who sat with the evaluation data three weeks after the launch and felt the particular grief of good work that met a bad implementation. Thomas, who received applause for a report and went back to his desk to think, for the first time, about what evaluation was actually for.

Evan and Francis, who argued productively across a table about which problem to solve first, and were both right. Angela, who learned about blood pressure monitors, and rebuilt her review process before the next project started. Jerome, who prepared for scenarios he never needed and was present for the one he did. Nadia, who said to the divisional head: "That's not what you asked for. I think it's what you need." Claudette, who opened a spreadsheet and started marking programs. Sinclair and Peter, who were not in the room when the decisions were being made, and whose absence was the failure. Lena, who found in an AI tool a partial substitute for the colleague she missed. And Silas, who made a presentation about change, and then understood what he had actually needed to do.

Each of them was doing the same thing from a different vantage point and with different tools: trying to understand what they were doing well, where they were falling short, and how to close the gap. That is the work. It has always been the work. AI has changed the tools available to it. It has not changed what "the work" means.

The Question That Remains

The question this book began with, Maria's question, on a Tuesday morning, was: if I can do this in thirty seconds, what exactly is my job now?

The answer the book has been building toward is not a defense of the instructional design role against the threat of AI. It is a clarification of what the role has always been, now that the tools have stripped away the parts that were never really the core of it.

Your job is not to produce objectives. Your job is to know which objectives are the right ones, at the right cognitive level, for these learners in this organizational context, in service of this performance outcome. AI can draft objectives. It cannot make that determination.

Your job is not to write scenarios. Your job is to know which scenarios will create genuine cognitive engagement for this audience, which will feel authentic to the work they actually do, and which will teach by implication what you intend them to teach. AI can generate scenarios. It cannot exercise that judgment.

Your job is not to deliver training. Your job is to understand what people need to be able to do, to determine whether training is the right way to develop that ability, to design the experience that will actually produce it, and to measure honestly whether it did. AI can accelerate parts of all of these. It cannot substitute for any of them.

Your job is to advocate for the learner, in the design process, in the stakeholder conversation, in the quality review, in the governance decision about which AI tools are used and how. The learner is the person who encounters what you build. They are the person for whom it either works or doesn't. They are the person whose competence, confidence, and professional development depend on the quality of your decisions. AI cannot advocate for them. You can.

A Note on Method

This book was written with AI assistance, as described in the Preface, and as demonstrated throughout. The research synthesis, the structural suggestions, the prompt frameworks in Appendix A, the first drafts of most chapters: all of it involved AI tools working alongside the human authorship that shaped, evaluated, revised, and takes responsibility for what ended up on the page. The final manuscript reflects the author's voice, the author's judgments, and the author's revisions applied to material that AI helped produce.

The demonstration is this: everything this book argues about the relationship between AI and human expertise in instructional design applies to the writing of a book about that relationship. The AI produced structure and content faster than a human working alone. The human judgment determined which structure was right, which content was accurate, which arguments were honest, which examples would resonate, which sentences needed to be rewritten from scratch, and which ideas the AI had generated that were not quite right in ways that took considerable thought to name.

The quality of this book is a function of that collaboration, and of which party held the editorial judgment and took the professional responsibility for the result. The AI was the drafting partner. The authorship is human. That is exactly the relationship this book argues for between AI and the instructional designers reading it.

A Final Word

The professions that survive technological disruption are the ones that understood clearly what the technology changed and what it did not. The ones that confused the tool with the work, that defined their professional identity around the tasks the technology was taking over rather than the judgment the technology depended on, contracted or disappeared.

Instructional design is not contracting. It is being asked to do something harder: to maintain the rigor and the humanity of good learning design while working with tools that operate at a pace and scale that would have seemed impossible a decade ago. To be the person who knows what AI is good for and what it isn't, who can use it without being used by it, who can move faster without losing the quality that made the work worth doing.

That is not an easy thing to be. It requires everything the previous generation of this work required, the learning theory, the design judgment, the stakeholder skill, the ethical commitment, the professional courage, and it requires those things in a context that is moving faster than any previous context the profession has navigated.

It is also, in the most precise sense of the word, important work. The systems that are coming: the adaptive platforms, the AI tutors, the performance support environments woven into the fabric of how work happens, will shape how people learn and develop their capabilities across their working lives. The people who define the goals those systems pursue, who specify the pedagogical logic they enact, who hold the ethical standards they are governed by, and who advocate for the learners they serve will determine whether those systems produce genuine human development or the appearance of it.

Those people are instructional designers, learning experience designers, performance consultants, learning technologists, facilitators, L&D managers, and all the adjacent practitioners who make this ecosystem work. They are the people this book was written for.

The question the profession faces is not whether it will survive the arrival of AI in learning. It will. The question is whether it will lead the development of AI in learning, shaping these systems toward genuine human benefit, or follow it, adapting to whatever the technology produces and hoping for the best.

This book is a case for leading. Not because it is more comfortable, it is not, but because it is the only version of this profession's future that is worth working toward.

INDIE

INvestigation-Design · Development · Implementation · Enhancement

A Field Guide to AI-Augmented Learning & Development

A Note on What Comes Next

The appendices that follow contain the practical infrastructure referenced throughout the book: the full INDIE prompt library, organized by phase and task, with rationale and known limitations for each prompt; a theoretical quick reference for the learning frameworks cited in the text; a role comparison matrix showing which INDIE phases and AI applications are most relevant to each role; an AI tool evaluation rubric; and a curated further reading list.

These appendices are designed to be used, not merely read. The prompt library is a starting point; the prompts that matter most for your practice are the ones you will develop, refine, and document from your own project experience. The further reading list is a map, not a syllabus. The role matrix is a navigation tool for practitioners who want to return to the chapters most relevant to their current situation.

The book ends here. The work continues. It has always continued. That is the nature of a profession that exists to help people learn in a world where what needs to be learned never stops changing.

Appendix A
The INDIE Prompt Library

This appendix contains every significant prompt from the book, organized by INDIE phase and task. Each entry includes the prompt itself, a note on what it is designed to surface, and a statement of its known limitations, where the prompt reliably underperforms and where expert judgment must compensate.

These prompts are starting points, not scripts. The most valuable prompt you will use is the one you develop from your own project experience and document in your team's shared library. This appendix gives you the foundation. Your practice builds from it.

How to Use This Library

- Copy and paste the prompt text directly. Replace all [BRACKETED] fields with specific information about your project, audience, and context. The quality of the output is directly proportional to the specificity of the context you provide.
- Read the "Known limitations" note before using any prompt in a high-stakes context. The limitations are not disclaimers. They are specific guidance about where to apply additional scrutiny.
- Prompts marked with a chapter number refer to the chapter where the prompt was introduced and the context in which its design rationale is explained.
- When a prompt produces an output that is significantly better or worse than expected, document what you changed and what happened. That documentation is your contribution to the team's shared learning.

Prompt Index by Phase

PHASE IN — Investigation-Design (IN-01 through IN-11)

IN-01 Pre-Interview Stakeholder Preparation · IN-02 Stakeholder Interview Synthesis · IN-03 Needs Analysis and Root Cause Classification · IN-04 Root Cause Hypothesis Testing · IN-05 Learner Persona Development · IN-06 LXD Empathy Research Synthesis · IN-07 Learning Objective Development · IN-08 Objective Quality Review · IN-09 Curriculum Architecture Design · IN-10 Performance Consulting Decision · IN-11 SME Knowledge Elicitation

PHASE D — Development (D-01 through D-14)

D-01 through D-09: Gagné Events 1–9 · D-10 Pedagogical Quality Review · D-11 Emotional Arc Review · D-12 Accessibility Audit · D-13 AI Voice Script Optimization · D-14 Rapid Prototype Content

PHASE I — Implementation (I-01 through I-09)

I-01 Deployment Planning · I-02 Pre-Session Communication · I-03 Manager Transfer Guide · I-04 Facilitator Preparation · I-05 Pilot Feedback Synthesis · I-06 Session Preparation · I-07 Post-Session Synthesis · I-08 Virtual/Hybrid Adaptation · I-09 Project Kickoff Package

PHASE E — Enhancement (E-01 through E-04)

E-01 Four-Level Analysis · E-02 Executive Enhancement Report · E-03 Dashboard Design · E-04 Learning-to-Performance Gap

CROSS-CUTTING (XC-01 through XC-05)

XC-01 Portfolio Health Audit · XC-02 Change Readiness Assessment · XC-03 INDIE Adoption Communication · XC-04 Quarterly Leadership Report · XC-05 Missing Colleague Review

GOVERNANCE (GOV-01 through GOV-03)

GOV-01 Technical Constraint Definition · GOV-02 AI Tool Evaluation Framework · GOV-03 SME Pre-Review Briefing

IN PHASE *INvestigation-Design*

IN-01 Pre-Interview Stakeholder Preparation *→ Chapter 5*

Designed to surface: A preliminary gap hypothesis and targeted interview questions before the first stakeholder meeting — replacing a blank intake form with a working analytical position.

Prompt

I am preparing to meet with a senior stakeholder about a potential training initiative. Here is what I know so far:

Organization/department: [DESCRIBE]
Stated problem: [DESCRIBE WHAT WAS COMMUNICATED]
Available background: [PASTE ANY RELEVANT DOCS OR SUMMARIES]
Stakeholder role: [TITLE AND APPROXIMATE SCOPE]
Known constraints: [TIME, BUDGET, AUDIENCE SIZE, ETC.]

Please:

1. Generate a preliminary hypothesis about the actual performance gap (distinguish between training-addressable and non-training causes)
2. Identify the 5-7 most important questions to ask in this meeting
3. Flag any topics I should approach carefully given the organizational context

4. Note what information is missing that would significantly change the analysis

Frame questions using Knowles' adult learning principles: focus on performance problems and immediate application, not training topics.

Known limitations: *AI cannot know your organization's political dynamics, prior relationship history with this stakeholder, or the informal context that shapes what questions are safe to ask. Revise the question list based on your own knowledge of the relationship before the meeting.*

IN-02 Stakeholder Interview Synthesis *→ Chapter 5*

Designed to surface: Recurring themes, stakeholder tensions, and the gap between stated needs and actual performance problems from interview notes or transcripts.

Prompt

I have completed [NUMBER] stakeholder interviews about [TRAINING TOPIC/NEED].
Here are my notes/transcripts:

[PASTE NOTES OR TRANSCRIPTS]

Please analyze these and provide:

1. The 4-6 most significant themes across all conversations
2. Points of stakeholder alignment (where perspectives agree)
3. Points of tension or disagreement (where perspectives conflict)
4. The gap between what stakeholders say they want and what the data suggests they actually need
5. Unanswered questions that require follow-up
6. Your assessment of whether a training intervention is the appropriate response, or whether other interventions should be considered

Flag any statements that suggest the real problem may be different from the stated problem. Those dissonances are often where the real diagnosis lives.

Known limitations: *AI will identify explicit themes reliably. It cannot detect the politically significant subtext, what was conspicuously not said, which stakeholder's apparent agreement was performative, or what the power dynamics in the room meant for the data quality. These observations require the analyst's judgment.*

IN-03 Needs Analysis and Root Cause Classification *→ Chapter 5*

Designed to surface: A structured Gilbert BEM analysis that separates training-addressable causes from environmental, motivational, and systemic ones before any design decisions are made.

Prompt

I am conducting a needs analysis for [JOB ROLE/DEPARTMENT] at [ORGANIZATION TYPE].
Here is the available performance data:

Performance metrics: [INSERT DATA]
Error or quality patterns: [INSERT DATA]
Stakeholder descriptions of the problem: [SUMMARIZE]
Observation notes: [IF AVAILABLE]

Using Gilbert's Behavior Engineering Model, analyze this data to:

1. Classify potential causes across six categories:
   Environmental: Data/Information | Resources/Tools | Incentives/Consequences
   Person: Knowledge/Skills | Capacity | Motives/Preferences
2. Rate the strength of evidence for each category (Strong/Moderate/Weak/Absent)
3. Identify the 2-3 most likely primary causes with supporting data
4. Using Piaget's schema theory, assess whether the gap reflects:
   - Missing foundational schemas
   - Outdated schemas needing restructuring
   - Extension of existing schemas
5. Reach a preliminary conclusion: Is a training intervention appropriate?
   If yes, for which specific knowledge/skill gaps?

Flag where data is insufficient to reach a confident conclusion.

Known limitations: *AI can classify causes from data you provide but cannot assess organizational context, management culture, or the informal power dynamics that often explain why incentive or feedback problems persist despite being visible. The classification is a starting hypothesis, not a diagnosis.*

IN-04 Root Cause Hypothesis Testing → *Chapter 12*

Designed to surface: Disconfirming evidence and alternative hypotheses, the discipline that separates rigorous diagnosis from motivated reasoning.

Prompt

I have formed a preliminary root cause hypothesis for a performance gap in [ROLE/DEPARTMENT]: [DESCRIBE YOUR HYPOTHESIS].

Current evidence supporting this hypothesis: [LIST KEY EVIDENCE]
Evidence that could challenge this hypothesis: [LIST IF ANY]

Please help me test this hypothesis rigorously by:

1. Generating 5-7 questions specifically designed to DISCONFIRM the hypothesis (not to confirm it)
2. Identifying 3 alternative hypotheses the current evidence could support

equally, and what additional data would distinguish between them
3. Generating the 'if this is true, then we should see' chain:
what observable evidence would confirm this cause is primary?
What would be present that would be absent if a different cause were primary?
4. Identifying the stakeholder whose perspective is most needed and
least represented in the current analysis

Goal: the hypothesis that best explains all the evidence, not the one
that is easiest to defend.

Known limitations: *AI generates alternative hypotheses that are logically consistent with the data provided. It cannot generate the hypothesis that requires organizational knowledge you haven't shared. Always add at least one alternative hypothesis based on your own contextual knowledge before testing.*

IN-05 Learner Persona Development → *Chapter 5*

Designed to surface: Detailed personas capturing not just demographics and skill levels but the emotional profile, professional identity stakes, and ZPD estimates that drive design decisions.

Prompt

I need to develop learner personas for [TRAINING PROGRAM] targeting [JOB ROLE].
Here is the available data on this audience:

Demographics: [EXPERIENCE RANGE, EDUCATIONAL BACKGROUND, ETC.]
Current skill/competency levels: [ASSESSMENT OR OBSERVATION DATA]
Job challenges and pain points: [FROM INTERVIEWS OR SURVEYS]
Self-reported learning preferences: [IF AVAILABLE]
Practical constraints: [TIME, LOCATION, TECHNOLOGY ACCESS, ETC.]

Please create 3-4 distinct learner personas. For each include:
- Name, role, and experience level
- Current knowledge and skills relevant to this training
- Primary job challenges this training should address
- Motivations for engaging with the learning (WIIFM)
- Barriers to engagement and learning transfer
- Preferred learning modalities and formats
- Vygotsky's ZPD: what they can do independently vs. with support

Design personas to represent meaningfully different points in the audience distribution, not just demographic variation.

Note: These require validation against actual learner input before they drive design decisions.

Known limitations: *AI defaults to abstract professional archetypes without organizational specificity. Revised personas based on actual learner conversation are significantly more valuable than unrevised AI drafts. Always validate at least one persona against a real interview before using the set to drive design.*

IN-06 LXD Empathy Research Synthesis → *Chapter 9*

Designed to surface: The emotional profile, friction points, and delight opportunities that pure cognitive needs analysis misses — the learner experience data that changes design decisions.

Prompt

I have conducted empathy research with [NUMBER] learners from [AUDIENCE] for [PROGRAM TOPIC]. Here are my research notes:

Interview/observation notes: [PASTE OR SUMMARIZE]
Key quotes revealing emotional or experiential dimensions: [PASTE]
Observed friction points (confusion, frustration, avoidance): [DESCRIBE]

Please synthesize this research to produce:

1. An emotional profile of this audience's relationship with this topic:
   - Default emotional state when approaching this content
   - Sources of professional identity investment or threat
   - Existing associations with how this topic has been trained before
2. A friction point map: specific moments in a learning experience where this audience is most likely to disengage or experience confusion
3. A delight opportunity map: moments where genuine recognition, competence, or connection is possible
4. Design implications: 3-5 specific design decisions this empathy data should change compared to what a purely cognitive analysis would recommend

Known limitations: *AI is pattern-matching your notes, not synthesizing emotional experience directly. The most important empathy insights — the thing said off-the-record, the body language during the observation session, the sentence that revealed an assumption neither party had named — are the ones most dependent on the researcher's presence and judgment to capture.*

IN-07 Learning Objective Development → *Chapter 5*

Designed to surface: Mager-format objectives at the correct Bloom's level for the performance required, distributed logically from foundational to complex.

Prompt

I am writing learning objectives for [COURSE/PROGRAM] targeting [AUDIENCE].

Performance gap identified: [DESCRIBE FROM NEEDS ANALYSIS]
Job requirements: [DESCRIBE WHAT LEARNERS MUST DO ON THE JOB]
Current competency level: [FROM LEARNER ANALYSIS]
Business goals this training serves: [DESCRIBE]
Available learning time: [HOURS/DAYS]

Generate 5-8 learning objectives using Mager's three-part format:
'Given [condition], the learner will [performance verb] to [criterion].'

For each objective:
- Specify the Bloom's Taxonomy level
- Confirm the cognitive level matches the actual job requirement
- Propose the most appropriate assessment approach for this objective

Organize objectives in a logical progression from foundational to complex.
Flag any objective where the Bloom's level required by the job exceeds what can be reasonably achieved in the available learning time.

Known limitations: *AI produces objectives that are structurally compliant (Performance + Condition + Criterion) but may be at the wrong Bloom's level, address the wrong gap, or use criterion standards that are not measurable in the actual assessment context. Every objective requires review against the actual job requirement, not just against the format.*

IN-08 Objective Quality Review → *Chapter 5*

Designed to surface: A systematic audit of existing objectives against Mager's criteria and Bloom's taxonomy alignment — catching format compliance that masks substantive problems.

Prompt

Please review the following learning objectives for [COURSE TITLE]:

[PASTE OBJECTIVES]

For each objective, evaluate:
1. Mager compliance: Does it specify Performance, Condition, and Criterion? If not, what is missing?
2. Bloom's alignment: What cognitive level does this target? Is that level appropriate for the job requirement it serves?
3. Measurability: Can performance be observed and assessed against the criterion?
4. Adult learning alignment: Does this connect to a real performance problem, or does it reflect a content topic?

Provide a revised version of any objective that fails these criteria, with a brief explanation of what was changed and why.

Known limitations: *AI evaluates objectives against the format and the logical coherence of the language. It cannot evaluate whether the objective addresses the right performance gap, that requires knowing what the needs analysis actually found. Use this prompt after verifying that the objectives address the right gaps, not as a substitute for that verification.*

IN-09 Curriculum Architecture Design → *Chapter 5*

Designed to surface: A scaffolded learning pathway that sequences content from supported to independent performance, with modality recommendations and cognitive load distribution.

Prompt

I need to design curriculum architecture for [PROGRAM TITLE].

Validated learning objectives: [PASTE OBJECTIVES]
Learner personas: [SUMMARIZE KEY CHARACTERISTICS]
Available learning time: [TOTAL HOURS, FORMAT]
Available modalities: [e.g., eLearning, ILT, Virtual ILT, Job Aid, Coaching]
LMS and technology constraints: [DESCRIBE]
Transfer environment: [DESCRIBE JOB CONTEXT WHERE SKILLS WILL BE APPLIED]

Using Vygotsky's scaffolding principles, design a curriculum architecture that includes:

1. Module/unit structure with title and primary objectives for each
2. Recommended sequencing with rationale
3. Recommended modality for each module with justification
4. Scaffolding plan: what support structures are provided at each stage and when they are reduced
5. Assessment checkpoint placement
6. Estimated time allocation per module
7. Prerequisites: what must be true about learners before each module

Flag any place where cognitive load may exceed what the target audience can manage, and suggest a mitigation.

Known limitations: *AI produces sequencing that is logically coherent but may not reflect the organizational constraints, learner relationship dynamics, or practical delivery realities that would reshape the architecture. The most consequential architectural decisions — what to cut when time is compressed, what scaffolding to remove when budget is reduced — remain judgment calls.*

IN-10 Performance Consulting: Is Training the Right Solution? *→ Chapter 9*

Designed to surface: A Gilbert BEM-informed assessment of whether a training intervention is appropriate before any design resources are committed.

Prompt

I have received a request for training on [TOPIC] for [AUDIENCE].
Before designing anything, I need to assess whether training is the right intervention.

Stated problem: [DESCRIBE]
Current performance: [DESCRIBE IF KNOWN]
Desired performance: [DESCRIBE IF KNOWN]
Available context: [PASTE ANY RELEVANT DOCS, DATA, OR NOTES]

Analyze this request against Gilbert's Behavior Engineering Model,
classifying potential causes across six categories:

1. Data/information: Do people know what is expected and how they are doing?
2. Resources/tools: Do people have what they need to perform?
3. Incentives/consequences: Are the right behaviors reinforced?
4. Knowledge/skills: Do people know HOW to perform?
5. Capacity: Are people physically/cognitively able to perform?
6. Motives/preferences: Do people want to perform?

Based on this analysis:

- What proportion of the gap is likely training-addressable (Category 4)?
- What non-training interventions should be considered alongside or instead?
- What additional information would significantly change this assessment?
- What should I say to the stakeholder if training is not the primary solution?

Known limitations: *AI can classify causes from the context provided but cannot assess organizational politics, management willingness to address non-training causes, or the organizational history that explains why a feedback problem has persisted for two years despite being visible. The most important performance consulting judgment, whether the organization will act on a non-training recommendation, cannot be made from data.*

IN-11 SME Knowledge Elicitation Interview Design → *Chapter 14*

Designed to surface: A structured interview guide that surfaces conditional and contextual knowledge — the edge cases, judgment points, and organizational specificity that documents cannot provide.

Prompt

```
I am preparing to interview [JOB TITLE / SME] about [TOPIC AREA]
for a learning program targeting [AUDIENCE].

I have already gathered: [DESCRIBE AVAILABLE DOCUMENTS, PROCEDURES,
  EXISTING TRAINING CONTENT]

AI has produced draft content for the declarative and procedural components.
The gaps I need this SME to fill are in conditional and contextual knowledge.

Please design a 60-minute knowledge elicitation interview that:
  1. Opens by explaining the SME's specific contribution (5 min)
  2. Elicits conditional knowledge (25 min):
      - When does the standard procedure not apply?
      - What would a novice do that an experienced practitioner wouldn't?
      - Walk me through a recent case that was harder than it looked.
  3. Elicits contextual knowledge (20 min):
      - How does this actually work here vs. how it's documented to work?
      - What has changed recently that someone new wouldn't know?
  4. Closes with scenario validation (10 min):
      - Review the scenario I've drafted. What's wrong with it?

Generate specific questions for each section, not just categories.
```

Known limitations: *AI generates questions that are generically appropriate for the domain. The questions that unlock the most valuable conditional knowledge are almost always specific to this SME's experience, the case they mentioned in passing during a previous conversation, the edge case their team encountered last month. Listen for those leads and follow them rather than working through the guide mechanically.*

D PHASE *Development*

D-01 Event 1: Gain Attention — Opening Scenario Development *→ Chapter 6*

Designed to surface: Multiple compelling opening scenarios that create genuine cognitive readiness through authentic workplace consequences, not just topic announcements.

Prompt

```
Create 3 distinct opening scenarios for a module on [TOPIC] for [AUDIENCE].
Each scenario should:
  - Connect directly to a real consequence of poor performance in this area
  - Be emotionally plausible for the [JOB ROLE] context
  - Create cognitive dissonance or concern that motivates engagement
  - Be completable in 2-3 minutes (eLearning) or 5 minutes (ILT)
  - Lead naturally into the statement of learning objectives

For each scenario, provide:
  - Setting and characters
  - The situation and the decision point
  - The consequence that creates the hook
  - The transition question that bridges to the objectives

Specify the industry/organizational context: [DESCRIBE].
Avoid scenarios that place blame on individual learners; focus on systemic
or situational factors that any reasonable person might encounter.
```

Known limitations: *AI scenarios are often emotionally safe — the consequence is visible but not quite sharp enough to create real concern. Review each scenario for whether a learner would actually feel the urgency, not just understand it intellectually. Also audit for demographic defaults: AI scenario characters cluster around a narrow demographic without explicit specification.*

D-02 Event 2: Learner-Facing Objective Presentation *→ Chapter 6*

Designed to surface: A translation of Mager-format design objectives into learner-facing language that connects performance to job outcome and answers the implicit WIIFM question.

Prompt

```
I have the following Mager-format learning objectives for a module on [TOPIC]:

  [PASTE OBJECTIVES]

Please create a learner-facing presentation of these objectives that:
```

- Translates each objective into plain language a [JOB ROLE] would recognize
- Connects each objective explicitly to a job performance outcome they care about
- Answers the implicit question: "Why does this matter to me, specifically?"
- Takes no more than 3 minutes to read or view
- Is appropriate for [DELIVERY FORMAT: eLearning / ILT / blended]

Additionally, propose one interactive alternative to a static objective slide: a brief activity that reveals or reinforces the objectives through learner engagement rather than passive reading.

Known limitations: *AI translations tend toward corporate motivation language ('By the end of this module, you will be empowered to...'). Review for genuine specificity: does this version of the objective connect to something this audience actually cares about, or does it describe competence in the abstract?*

D-03 Event 3: Recall Activation Activities → *Chapter 6*

Designed to surface: Activities that activate prior knowledge explicitly and surface likely misconceptions before new content is introduced — building the cognitive bridge rather than assuming it.

Prompt

I am designing Event 3 for a module on [TOPIC] for [AUDIENCE].

From my learner analysis, I know:
- Current knowledge level: [DESCRIBE]
- Likely correct prior knowledge: [DESCRIBE]
- Likely misconceptions or outdated schemas: [DESCRIBE IF KNOWN]
- Delivery format: [eLearning / ILT / blended]

Please design 2-3 recall activity options that:
- Begin from what learners already know, not from what they don't
- Surface any common misconceptions new content will need to address
- Connect prior experience to upcoming content explicitly
- Take no more than 5 minutes

For each option, note whether it is more appropriate for novice or experienced audiences, and flag any risk of the activity feeling condescending to practitioners with significant prior experience.

Known limitations: *AI defaults to 'what do you know about X?' quiz formats that feel like pre-tests rather than activation. The most effective recall activities for experienced learners often begin from a real situation they've encountered, not from a question about their knowledge. Review for whether the activity honors prior experience or implicitly frames the learner as deficient.*

D-04 Event 4: Chunked Content Outline → *Chapter 6*

Designed to surface: A content structure organized by working memory capacity rather than subject matter logic — maximum five key concepts per unit, sequenced from familiar to unfamiliar.

Prompt

```
I need to develop the content presentation for a module on [TOPIC].

  Learning objective this content serves: [PASTE OBJECTIVE]
  Target audience: [DESCRIBE: role, experience level, context]
  Available time for this section: [MINUTES]
  Source material: [PASTE OR DESCRIBE]

Please produce a chunked content outline that:
  1. Identifies 3-5 key concepts (no more, prioritize ruthlessly)
  2. For each concept, provides:
     - A plain-language explanation appropriate for this audience
     - One concrete example from the [INDUSTRY/ROLE] context
     - One common error or misconception to address
     - A cognitive pause indicator (where the learner should stop
       and process before moving on)
  3. Sequences concepts from familiar-to-unfamiliar and concrete-to-abstract
  4. Flags any concept where intrinsic complexity is high enough to warrant
     additional scaffolding or prerequisite review

Apply Sweller's cognitive load principles: eliminate elaboration that does
not serve the learning objective.
```

Known limitations: *AI produces comprehensive outlines that respect the five-concept guideline in structure but not always in density, each "concept" may contain three or four distinct ideas without the prompt flagging it. Apply the five-concept audit yourself at the paragraph level, not just the heading level.*

D-05 Event 5: Learning Guidance and Job Aids → *Chapter 6*

Designed to surface: Worked examples that narrate expert reasoning (not just expert outcomes) and job aids designed for use under realistic performance conditions.

Prompt

```
I need to develop learning guidance for [TOPIC] following content presentation.

  Key concepts just presented: [LIST THE 3-5 CONCEPTS FROM EVENT 4]
  Target audience: [DESCRIBE]
  Most common application context: [WHERE WILL LEARNERS USE THIS?]
  Known points of confusion or difficulty: [DESCRIBE IF KNOWN]
```

Please develop the following guidance tools:

1. One worked example showing an expert applying [SPECIFIC CONCEPT] in a realistic [JOB ROLE] situation, with narrated decision points that make the expert reasoning visible — not just the outcome

2. One job aid (decision tree OR quick reference OR checklist) that a learner could use during actual job performance, not just during training

3. Three expert tips that compress practitioner wisdom about common pitfalls, edge cases, or efficiency shortcuts in this domain

For the worked example, show the reasoning process, not just the outcome. For the job aid, optimize for use under realistic time pressure.

Known limitations: *Worked examples are AI's most reliable Event 5 output structurally, but the expert reasoning narrated is often generic. The specific decision points that matter for this audience in this organizational context — what makes the hard case hard — require SME input that is not available to the AI from a prompt alone.*

D-06 Event 6: Graduated Practice Activities → *Chapter 6*

Designed to surface: Three complexity levels of scenario-based practice that match the cognitive operation the objective requires and accommodate the full ZPD range of the audience.

Prompt

I need to design practice activities for [SKILL/TOPIC] for [AUDIENCE].

Learning objective: [PASTE OBJECTIVE — include Bloom's level]
Job context where skill will be applied: [DESCRIBE]
Delivery format: [eLearning / ILT / blended]
Known audience characteristics: [RELEVANT PERSONA DETAILS]

Please design 3 practice activities at graduated complexity levels:

Level 1 (Foundational): High scaffolding, clear context, limited variables
Level 2 (Applied): Moderate scaffolding, realistic ambiguity, job-authentic context
Level 3 (Integrated): Minimal scaffolding, competing priorities, requires judgment

For each activity:
- Scenario setup (setting, characters, situation)
- The task the learner must complete

- The cognitive operation required (must match Bloom's level)
- Success criteria
- Time allocation

For Level 3: include at least one plausible distractor or red herring that requires judgment rather than procedure application.

Known limitations: *AI practice scenarios are often too clean, the situation telegraphs the correct response more clearly than a real workplace situation would. Review Level 2 and Level 3 scenarios specifically for whether a learner who knows the right answer might still miss it, which is the condition that produces genuine learning.*

D-07 Event 7: Instructional Feedback Scripts *→ Chapter 6*

Designed to surface: Feedback that explains correct reasoning, addresses incorrect reasoning specifically, and redirects without evaluating the learner — feedback that teaches rather than merely judges.

Prompt

I need to develop feedback scripts for the following practice activity:

Practice scenario: [DESCRIBE THE SCENARIO]
Correct response: [DESCRIBE WHAT CORRECT PERFORMANCE LOOKS LIKE]
Common incorrect responses: [LIST IF KNOWN]

Please develop feedback scripts for:

1. Correct response feedback:
   - Confirm the correct reasoning (not just the correct answer)
   - Reinforce the principle that was applied
   - Extend understanding with one insight the learner may not have considered

2. For each plausible incorrect response:
   - Acknowledge what was reasonable about the incorrect reasoning
   - Explain specifically what was wrong and why
   - Redirect to the correct principle without simply stating the answer
   - Frame the correction in terms of performance improvement, not evaluation of the learner

3. A self-assessment prompt the learner can use after receiving feedback to consolidate the learning

Tone: Collegial, non-condescending. Assumes professional adults who are capable of learning from mistakes without being diminished by them.

Known limitations: *AI feedback is structurally sound but often too comprehensive — it explains everything at once rather than identifying the one most important thing to correct. Review for parsimony: what is the single most consequential thing this learner needs to understand from this feedback, and does the script lead with that?*

D-08 Event 8: Assessment Item Development → *Chapter 6*

Designed to surface: Scenario-based assessment items across multiple formats, with plausible distractors that represent real misconceptions rather than obviously wrong answers.

Prompt

I need to develop summative assessment items for [MODULE TITLE].

Validated learning objectives: [PASTE ALL OBJECTIVES FOR THIS MODULE]
Target audience: [DESCRIBE]
Assessment format constraints: [eLearning quiz, observed performance, etc.]
Available assessment time: [MINUTES]

Please develop a balanced assessment that:

1. Includes at least one item per learning objective
2. Uses the following item types:
   - Scenario-based multiple choice (application level or higher)
   - Case analysis (analysis or evaluation level)
   - Performance checklist (for procedural objectives)
   - Reflection prompt (for objectives involving professional judgment)

For each item:

- Identify the objective it assesses
- Confirm the Bloom's level matches the objective
- Provide correct answer and rationale
- For multiple choice: provide 3 plausible distractors with the reasoning error each distractor represents

Flag any objective that cannot be adequately assessed in the available format or time, with a recommendation for an alternative assessment approach.

Known limitations: *AI distractors are the most unreliable output in assessment generation — they are often too obviously wrong or represent errors that real learners in this domain don't actually make. Have a domain SME review distractor quality specifically, not just content accuracy.*

D-09 Event 9: Retention and Transfer Design → *Chapter 6*

Designed to surface: A spaced retrieval schedule, performance support for the job context, and an application assignment that extends the learning into actual workplace practice.

Prompt

```
I need to design a retention and transfer strategy for [MODULE/PROGRAM TITLE].

 Learning objectives: [PASTE OBJECTIVES]
 Job context: [DESCRIBE WHERE AND HOW SKILLS WILL BE APPLIED]
 Transfer environment: [DESCRIBE SUPPORTS AND BARRIERS TO TRANSFER]
 Available follow-up touchpoints: [LMS push, email, manager check-in, etc.]
 Manager involvement available: [YES / NO / LIMITED]

Please develop a retention and transfer plan that includes:

 1. Spaced retrieval schedule:
    - 3 brief retrieval prompts at Day 3, Day 14, Day 30
    - Each prompt should take 2-3 minutes and require active recall

 2. Performance support tool:
    - Designed for use during actual job performance
    - Optimized for realistic conditions of use

 3. Application assignment:
    - A structured task the learner completes in their actual job
      within 2 weeks of training
    - Includes a brief reflection connecting the experience back
      to the learning objectives

 4. If manager involvement is available: a manager follow-up guide
    with 3 coaching questions for a 10-minute post-training conversation
```

Known limitations: *AI transfer strategies are generically sound but do not account for the specific transfer barriers in your organization — the manager who won't reinforce the learning, the workflow that makes the new approach impractical, the culture that rewards the old behavior. These barriers require organizational knowledge the AI doesn't have.*

D-10 Pedagogical Quality Review of AI-Generated Content → *Chapter 9*

Designed to surface: A systematic audit of AI-generated content against the objective, the required Bloom's level, cognitive load standards, and organizational authenticity — catching the 'organized but not instructional' failure mode.

Prompt

```
I need to conduct a pedagogical quality review of the following
AI-generated content for [MODULE TITLE]:

  [PASTE AI-GENERATED CONTENT]

  Validated learning objective this content serves: [PASTE OBJECTIVE]
  Target audience: [DESCRIBE]
  Bloom's level required: [SPECIFY]

 Please evaluate this content against the following criteria:
  1. Objective alignment: Does this directly serve the stated objective,
     or address adjacent content that is accurate but not required?
  2. Cognitive level: Does the content develop the cognitive operation
     the objective requires, or a lower-level operation?
  3. Cognitive load: Does this exceed working memory capacity?
     Identify any sections with more than 5 distinct concepts.
  4. Schema connection: Does this connect to prior knowledge the audience has?
     Or does it assume foundational knowledge that may not be present?
  5. Example quality: Are examples specific to the [INDUSTRY/ROLE] context,
     or generic enough to feel inauthentic?
  6. Extraneous content: What could be removed without reducing effectiveness?

 Provide a revised version of any section that fails these criteria,
 with brief annotations explaining each change.
```

Known limitations: *This prompt asks AI to evaluate AI output, which it can do usefully for structural and format criteria but less reliably for the organizational authenticity of examples. An AI cannot know whether a scenario feels authentic to your specific learner population. The example quality criterion requires your judgment.*

D-11 Emotional Arc Design Review → *Chapter 9*

Designed to surface: An evaluation of the learner's target emotional state at each stage of the program and specific design interventions that produce and sustain that state.

Prompt

```
I need to design and review the emotional arc for [PROGRAM/MODULE TITLE].
```

Target audience: [DESCRIBE — include emotional profile from empathy research]
Program structure: [DESCRIBE MODULES, ACTIVITIES, SEQUENCE]
Delivery format: [eLearning / ILT / blended]
Known audience emotional starting point: [FROM EMPATHY RESEARCH]

Part 1: Arc Design
For each stage of the program, specify:
- The target emotional state at this stage
- The specific design element(s) responsible for producing it
- The risk: what might produce the wrong emotional state, and how the design mitigates it

Part 2: Arc Review (if content already drafted)
Please review the following content [PASTE CONTENT] and identify:
- Moments where the emotional arc is likely to hold
- Moments where it is likely to break
- Specific language or structural changes that would restore the arc at identified break points

Focus on the learner's emotional experience, not the content's logical organization. Those are different things.

Known limitations: *AI can identify structural risk points in an emotional arc — places where the design logic predicts an emotional break — but cannot predict how a specific audience in a specific organizational context will actually respond. The pilot is the only reliable test of whether the designed arc matches the experienced arc.*

D-12 Accessibility Audit of Learning Content → *Chapter 9*

Designed to surface: A criterion-referenced review of content against WCAG 2.1 AA covering contrast, captions, alt text, keyboard navigation, reading order, language clarity, and motion.

Prompt

I need to conduct an accessibility audit of the following learning content before it moves to production:

Content type: [eLearning module / video / job aid / facilitator guide / etc.]
Content: [PASTE TEXT CONTENT, DESCRIBE VISUAL/INTERACTIVE ELEMENTS]
Authoring tool: [e.g., Articulate Storyline, Rise, Adobe Captivate]
Target accessibility standard: [WCAG 2.1 AA / Section 508 / other]

Please audit against the following criteria:
1. Color and contrast: Flag combinations that may fail 4.5:1 ratio,

or meaning carried by color alone
2. Language clarity: Flag jargon, passive constructions, or reading level issues for the specified audience
3. Alt text gaps: Evaluate whether descriptions convey informational content
4. Caption quality: If captions are present, identify accuracy errors
5. Keyboard accessibility: Identify elements requiring mouse-only interaction
6. Cognitive load: Flag sections where density may create barriers for learners with cognitive disabilities

For each issue: location, WCAG criterion affected, severity (critical/major/minor), and specific remediation.

Known limitations: *AI can audit text-based content reliably. It cannot audit visual contrast ratios without seeing the actual colors, cannot evaluate keyboard navigation behavior without interacting with the built module, and cannot assess caption accuracy against audio. Technical accessibility testing requires tool-based verification (e.g., WAVE, axe) in addition to this prompt.*

D-13 AI Voice Script Optimization *→ Chapter 10*

Designed to surface: A script revised for natural AI voice delivery: sentences broken for pacing, emphasis markers inserted, emotional register flags identified for potential human re-recording.

Prompt

I have the following narration script for AI voice synthesis:

[PASTE NARRATION SCRIPT]

Voice style target: [warm and collegial / authoritative and clear / etc.]
Content type: [PROCEDURAL / EXPLANATORY / SCENARIO-BASED]
AI voice tool: [e.g., ElevenLabs, Murf, Descript, Speechify]

Please review and optimize this script for AI voice delivery:
1. Sentence length: Break sentences longer than 20 words into two.
2. Emphasis markers: Insert [EMPHASIS] before words carrying instructional weight.
   Flag where AI default emphasis may conflict with instructional intent.
3. Pause markers: Insert [PAUSE] where a human narrator would breathe or where the listener needs processing time. Minimum: one per paragraph.
4. Emotional register flags: Mark any sentence where content requires gravity, warmth, or urgency that AI voice is unlikely to produce. Recommend human re-recording for those sections.
5. Jargon and acronyms: Flag terms AI voice may mispronounce and provide phonetic alternatives.

Return the optimized script with all markers in place.

Known limitations: *Even with an optimized script, AI voice quality is tool-specific and changes with each synthesis run. Always listen to the full synthesized output before approving for production — the script optimization reduces but does not eliminate AI voice failure modes.*

D-14 Rapid Prototype Content Generation → *Chapter 10*

Designed to surface: Representative placeholder content for stakeholder design validation — clearly marked as draft, structured to test interaction design without requiring production-quality content.

Prompt

```
I am building a rapid prototype for stakeholder review of [MODULE TITLE].

  Module topic: [DESCRIBE]
  Target audience: [DESCRIBE]
  Primary learning objective: [PASTE OBJECTIVE]
  Prototype scope: [NUMBER] slides covering [DESCRIBE COVERAGE]
  Key interaction to prototype: [DESCRIBE THE MAIN INTERACTION TYPE]

Please generate:
  1. On-screen text for each prototype slide (headline + 2-3 body points)
     Written at the correct reading level but labeled clearly as DRAFT CONTENT
  2. Narration script draft for 3 key slides (the most complex content)
  3. Scenario setup text for the prototype interaction:
       - Character and context (1 paragraph)
       - Decision prompt (1-2 sentences)
       - 3 response options with outcome descriptions
  4. Placeholder feedback for each response option

Flag any content where domain accuracy needs SME verification before
the prototype is shared with stakeholders outside the design team.

Note: This content is for design validation only. All content requires
expert review before production.
```

Known limitations: *Prototype content that looks too finished invites stakeholders to review production quality rather than design quality. If the AI-generated content is too polished, add explicit 'DRAFT — NOT FOR PRODUCTION' watermarks to every slide before sharing. The purpose of the prototype is to test the design, not to showcase the content.*

I PHASE *Implementation*

I-01 Deployment Planning → *Chapter 7*

Designed to surface: A phased rollout schedule with dependency mapping, risk log, and communications calendar — the organizational infrastructure that implementation requires.

Prompt

I need to plan the full rollout of [PROGRAM TITLE] following a successful pilot.

Total audience: [NUMBER] people across [NUMBER] locations/departments
Available delivery windows: [DESCRIBE CONSTRAINTS]
Facilitators available: [NUMBER AND EXPERIENCE LEVEL]
Technical infrastructure: [LMS, video platform, room requirements]
Rollout deadline: [DATE]
Program format: [LENGTH, MODALITY, COHORT SIZE]

Please produce:

1. A phased rollout schedule with buffer time for facilitator preparation
2. A dependency map identifying tasks that must precede each milestone
3. A risk log covering the 5 most likely deployment risks with probability assessment and mitigation for each
4. A communications calendar: what to send, to whom, and when, from four weeks before first session through two weeks after final session

Flag any constraint conflict that makes the current timeline unrealistic, with a specific recommendation for resolution.

Known limitations: *AI deployment plans are logically coherent but do not account for the organizational calendar dynamics that practitioners know — the month when the target department is always in planning cycles, the facilitator who is reliable on paper but consistently overcommits, the LMS administrator who needs three weeks for configuration changes despite formal SLAs of one. Layer these into the plan before presenting it.*

I-02 Pre-Session Participant Communication → *Chapter 7*

Designed to surface: A participant message that answers, "why should I care?" in 90 seconds, sets accurate expectations, and does not use L&D jargon.

Prompt

I need to write a pre-session message for participants in [PROGRAM TITLE].

Audience: [JOB ROLE AND DEPARTMENT]
Session format: [LENGTH, MODALITY, LOCATION/PLATFORM]
Primary learning objectives (in plain language): [DESCRIBE]

What participants should do to prepare: [DESCRIBE, OR 'none required']
What the program will and will not cover: [BRIEF SCOPE DESCRIPTION]
Organizational context (why this program, why now): [DESCRIBE]
Sender: [NAME AND ROLE]

Please draft a message that:
- Takes 90 seconds or less to read
- Answers the implicit question: 'Why should I care about this?'
- Sets accurate expectations without over-promising
- Is warm in tone without being performatively enthusiastic
- Does not use L&D jargon (no 'learning journey', 'upskilling', etc.)

Also draft a separate 3-sentence version for use as a calendar invite description.

Known limitations: *AI communications default to slightly more formal language and slightly more optimistic framing than most audiences respond to well. Read the draft aloud — if it sounds like a marketing communication rather than a colleague talking to colleagues, revise the tone down significantly.*

I-03 Manager Transfer Support Guide → *Chapter 7*

Designed to surface: A one-page manager guide with coaching questions, observable behavior indicators, and the single most important thing managers should avoid — designed for 5 minutes of manager time, not 30.

Prompt

I need to create a brief manager guide for supporting transfer after [PROGRAM TITLE].

What participants learned: [SUMMARIZE KEY COMPETENCIES IN PLAIN LANGUAGE]
How managers will see the learning appear in behavior: [DESCRIBE]
The 2-3 most common barriers to transfer in this environment: [DESCRIBE]
Time available for manager engagement: [REALISTIC ESTIMATE]

Please produce a one-page manager guide that includes:
1. What your team member experienced (2-3 sentences, no jargon)
2. What you might notice differently in their work in the next 30 days
3. Three coaching questions for a 5-minute follow-up conversation (questions that open dialogue, not questions that test recall)
4. One thing to avoid: the most common manager response that undermines transfer in this specific context

Tone: Collegial peer-to-peer. Managers are partners in transfer, not compliance officers.

Known limitations: *AI manager guides are often either too general ('support your team member's learning') or too prescriptive (detailed scripts that no manager will follow). The coaching questions are the most valuable output — review them specifically for whether they would actually open a productive conversation with a real manager, not just satisfy the checklist.*

I-04 Facilitator Preparation Guide → *Chapter 7*

Designed to surface: A preparation guide for facilitators who may not have designed the program they are delivering — covering the five most important audience characteristics, three critical facilitation moments, and likely Q&A.

Prompt

```
I need to create a facilitator preparation guide for [PROGRAM/SESSION TITLE].

  Session length: [DURATION]
  Participant profile: [DESCRIBE AUDIENCE]
  Facilitator background: [Subject matter expert? ID professional?
    First-time facilitator for this content?]
  Delivery format: [IN-PERSON / VIRTUAL / HYBRID]
  Materials provided: [LIST WHAT THE FACILITATOR WILL HAVE]

Please produce a preparation guide that includes:
  1. The 5 things most important to know about this audience before walking in
  2. The 3 moments in this session where facilitation skill matters most,
     and what skilled facilitation looks like at each
  3. The 3 most likely facilitation challenges and a specific response to each
  4. Questions participants are likely to ask, organized by section,
     with suggested responses
  5. Technology and logistics checklist (24 hours before, 30 minutes before,
     immediately before session start)

This guide is for the facilitator's use, not for participants.
Write in a direct, collegial voice — peer practitioner to peer practitioner.
```

Known limitations: *AI facilitation guides are thorough but often over-scripted — facilitators who try to follow them precisely produce the stilted delivery that over-preparation creates. Brief the facilitator that the guide is preparation material, not a script, and that the three critical moments (item 2) are the only sections worth memorizing.*

I-05 Rapid Pilot Feedback Synthesis → *Chapter 7*

Designed to surface: A categorized issue list and priority ranking from pilot feedback data — organized by design, delivery, and context issues, with specific revision recommendations for each high-priority finding.

Prompt

```
I need to synthesize pilot feedback for [PROGRAM TITLE] rapidly.

  Participant reaction survey responses: [PASTE QUANTITATIVE RATINGS
    AND OPEN-ENDED RESPONSES]
  Facilitator debrief notes: [PASTE OR SUMMARIZE]
  Observer field notes: [PASTE OR SUMMARIZE]
  Informal participant comments captured post-session: [PASTE IF AVAILABLE]
  Number of participants: [NUMBER]

Please produce:
  1. A themes analysis of the open-ended feedback (5-7 themes,
     frequency-weighted, with representative quotes for each)
  2. A categorized issue list organized by:
     - Design issues (content, structure, activities, assessments)
     - Delivery issues (facilitation, pacing, technology)
     - Context issues (scheduling, environment, audience preparation)
  3. A priority ranking of the top 5 issues to address before the next cohort,
     with a brief rationale for each ranking
  4. Specific revision recommendations for each high-priority design issue

Flag any finding that suggests the program objectives may not be the right
objectives for this audience.
```

Known limitations: *AI themes analysis is reliable for explicit feedback. It misses the implicit feedback — the participant who gave 4/5 on everything but wrote a single pointed observation in the open-ended box, the facilitator's body language in the debrief, the thing everyone mentioned casually that nobody put in the survey. Supplement with your own qualitative read before prioritizing.*

I-06 Session Preparation: Anticipating the Room → *Chapter 11*

Designed to surface: A structured facilitation preparation covering audience dynamics, hard moment anticipation, and multiple opening alternatives — the cognitive work that creates presence in the room.

Prompt

```
I am facilitating [SESSION TITLE] for [AUDIENCE DESCRIPTION].
Session length: [DURATION]. Format: [IN-PERSON / VIRTUAL / HYBRID].
Here is the session plan: [PASTE OR DESCRIBE SESSION PLAN]
Please help me prepare by generating:
```

1. Participant profile based on [ROLE/DEPARTMENT/CONTEXT]:
   - Likely prior experiences with this topic (positive and negative)
   - Skepticisms or resistances I should anticipate
   - Questions likely to come up beyond the planned content
   - Group dynamics if multiple departments or levels are present

2. Hard moment anticipation for each major segment:
   - Activity that might fall flat and why
   - Discussion that might go nowhere or everywhere
   - Transition that might lose the room

   For each challenge: three possible responses (redirect / name the dynamic / adapt the plan)

3. Opening alternatives:
   - Option A: Creates curiosity and safety
   - Option B: Creates urgency and relevance
   - Option C: Activates prior experience and honors expertise

4. Two closing synthesis approaches that help participants name what they will do differently as a result of today.

Known limitations: *AI preparation scenarios are logically plausible but don't know your specific audience. The most important preparation conversations — with a colleague who knows these participants, or with the program sponsor who knows what's politically live right now — cannot be replaced by this prompt. Use it after those conversations, not instead of them.*

I-07 Post-Session Synthesis Document → *Chapter 11*

Designed to surface: A group-specific synthesis of session insights for participant follow-up — not a generic summary of the content, but a record of what THIS group produced.

Prompt

I have just facilitated [SESSION TITLE] with [NUMBER] participants. Here are my notes from the session:

Key themes that emerged in discussion: [DESCRIBE]
Moments of insight or breakthrough: [DESCRIBE SPECIFIC MOMENTS]
Questions raised but not fully resolved: [LIST]
Activities that generated strong engagement: [DESCRIBE]
Commitments or intended actions participants mentioned: [LIST IF CAPTURED]

Please produce a post-session synthesis document for participants that:

1. Opens with 2 sentences framing what this group accomplished together (specific to THIS group, not generic to the topic)
2. Captures the 3-5 most significant insights that emerged from the group's discussion (attributed to the group, not the content)
3. Names the questions the group surfaced but didn't fully resolve, framed as productive tensions worth continued reflection
4. Connects the session's content to one concrete application opportunity for this specific audience
5. Closes with one question for individual reflection in the next 48 hours

Tone: Warm, specific, collegial. One page maximum.

Known limitations: *The value of this document is in its specificity to this group. AI will produce a structurally appropriate document from your notes, but the first and second items — framing what this group accomplished and attributing insights to the group — require your editorial judgment about what was most significant and how it should be named.*

I-08 Virtual and Hybrid Session Adaptation → *Chapter 11*

Designed to surface: A format adaptation that preserves pedagogical purpose across delivery modalities, with presence-building structures specifically designed to prevent the audience dynamic that hybrid creates.

Prompt

I need to adapt [SESSION TITLE] for virtual / hybrid delivery.
Original format: [IN-PERSON, DESCRIBE ORIGINAL DESIGN]
New format: [FULLY VIRTUAL / HYBRID — specify in-room vs. remote numbers]
Platform: [ZOOM / TEAMS / WEBEX / OTHER]
Session length: [DURATION]

Please redesign this session for the new format:

1. Format-specific challenges: What are the 3 most likely facilitation challenges specific to [virtual/hybrid] delivery of THIS content with THIS audience?

2. Activity adaptations: For each activity in the original session, propose a version that works in the new format while preserving the pedagogical purpose. Flag any activity that cannot be meaningfully adapted and should be replaced.

3. Presence-building structures: Design 2 specific structures that generate visible participation from remote participants and prevent the audience dynamic that hybrid facilitation tends to produce.

4. Technical pre-checks: List specific elements to verify 30 minutes before this session on [PLATFORM].

5. Recovery plans: For each likely technical failure, provide a facilitation response that keeps the group engaged.

Known limitations: *AI hybrid session designs often solve the technical problem (how to include remote participants in an activity) without addressing the social problem (how to ensure remote participants feel genuinely included rather than accommodated). Review each adaptation specifically for whether a remote participant would experience full participation or a graceful workaround.*

I-09 Project Kickoff Package → *Chapter 14*

Designed to surface: A charter, RACI matrix, risk log, communication plan, and scope boundary statement — the project management infrastructure that prevents the most common L&D project failures.

Prompt

I need to prepare the project kickoff package for [PROGRAM TITLE].
Project scope: [DESCRIBE]
Team: [LIST ROLES]
Stakeholders: [LIST WITH ROLES]
Timeline: [START DATE to TARGET COMPLETION]
Key risks already identified: [LIST]
Deliverable milestones: [LIST WITH TARGET DATES]

Please produce a kickoff package including:

1. Project charter (one page):
   Purpose, scope, success criteria, team roles and responsibilities, key milestones, and decision-making authority
2. RACI matrix for the six most critical project decisions
3. Risk log with the five highest-probability risks:
   Probability (H/M/L), Impact (H/M/L), mitigation action, owner
4. Communication plan:
   Who receives what updates, at what frequency, in what format
5. Scope boundary statement:
   Three specific things this project WILL deliver and three specific things it will NOT deliver

Known limitations: *AI scope boundary statements are the most reliably underspecified output in this prompt. The statement 'this project will not include facilitation training for managers' sounds specific but will not hold in a scope conversation. Specificity means naming the deliverable, the format, the review cycle, and the completion criteria, not just the category.*

E PHASE *Enhancement*

E-01 Kirkpatrick Four-Level Analysis → *Chapter 8*

Designed to surface: A cross-level analysis that connects evaluation data to specific design and delivery elements — not a data summary but a diagnostic that identifies what to change.

Prompt

```
I need to analyze Enhancement data for [PROGRAM TITLE] across all four
Kirkpatrick levels.

  Level 1 data: [PASTE QUANTITATIVE SCORES AND OPEN-ENDED RESPONSES]
  Level 2 data: [PASTE ASSESSMENT PERFORMANCE BY OBJECTIVE AND ITEM]
  Level 3 data: [PASTE TRANSFER SURVEY RESPONSES AND/OR OBSERVATION
DATA]
  Level 4 data: [PASTE BUSINESS METRIC DATA AND BASELINE]
  Number of participants: [NUMBER]
  Program objectives: [PASTE]
  Business metric this program was designed to influence: [DESCRIBE]

Please produce:
  1. An analysis at each Kirkpatrick level:
     - What the data shows (key patterns, not raw numbers)
     - What it suggests about program effectiveness
     - The specific design or delivery element most likely to explain the pattern
     - The recommended action
  2. Cross-level correlations:
     - Where Level 1 scores correlate or fail to correlate with Level 2
     - Where Level 2 achievement correlates or fails to correlate with Level 3
     - Any pattern suggesting the evaluation may be measuring the wrong things
  3. A priority-ranked list of 3 improvements with rationale based on
     impact potential and implementation effort
  4. A one-paragraph executive summary for a program sponsor with 90 seconds
```

Known limitations: *Level 4 correlation analysis is where this prompt is most limited: AI can identify correlations between program completion and metric change but cannot assess causation, control for confounding variables, or determine whether the timeline is sufficient for behavioral change to produce measurable results. Present Level 4 findings with explicit methodological caveats.*

E-02 Executive Enhancement Report → *Chapter 8*

Designed to surface: An audience-calibrated executive report that connects program performance to business outcomes, acknowledges gaps honestly, and closes with a specific decision request.

Prompt

I need to write an executive report on [PROGRAM TITLE] for [CHRO / VP / EXECUTIVE LEADERSHIP TEAM].

Program purpose: [ONE SENTENCE]
Total participants to date: [NUMBER]
Time period covered: [DATE RANGE]
Key findings from Kirkpatrick analysis: [SUMMARIZE]
Business metric movement: [DESCRIBE — INCLUDE BASELINE AND CURRENT]
Investment made: [APPROXIMATE COST IF AVAILABLE]
Key improvements made since launch: [LIST 2-3]

Please produce an executive report that:

1. Opens with one sentence: the program's business purpose and current status
2. Presents 3-4 key findings as headlines with one supporting sentence each
3. Connects competency outcomes to business metric outcomes explicitly
4. Acknowledges one area where results are not yet where they should be, and what is being done about it
5. Closes with a specific recommendation and requested decision
6. Fits on two pages maximum

Tone: Confident and data-grounded. Honest about gaps. Minimal L&D jargon.

Known limitations: *AI executive reports tend toward diplomatic balance — presenting strengths and gaps with equal weight, which produces reports that feel reassuring but make weak cases. Review specifically for whether the report builds a clear argument toward the requested decision, or whether it hedges that argument into neutrality.*

E-03 Enhancement Dashboard Design → *Chapter 8*

Designed to surface: A dashboard specification designed for decision support rather than reporting — with audience-differentiated views and alert thresholds that trigger specific actions.

Prompt

I need to design an Enhancement dashboard for [PROGRAM TITLE].
Program description: [BRIEF SUMMARY]
Primary stakeholders: [LIST BY ROLE]
Data sources available: [LMS / SURVEY TOOL / HRIS / PERFORMANCE SYSTEM]
Update frequency achievable: [REAL-TIME / DAILY / WEEKLY]
Business metric this program influences: [DESCRIBE]

Please design a dashboard specification that includes:
1. Designer view:
   - Completion rate by cohort and session
   - Level 2 assessment performance by objective
   - Level 1 satisfaction scores with open-ended themes
   - Level 3 transfer rate by competency
   - Flagged alerts: items falling below defined thresholds
2. Executive/sponsor view:
   - Program reach vs. target
   - Headline competency achievement rate
   - Level 4 business metric trend
   - One key insight and one recommended action
3. Facilitator view:
   - Their session-level scores vs. program average
   - Specific open-ended feedback from their sessions
   - Suggested adjustments based on session data
4. Alert thresholds with the recommended action for each

Known limitations: *AI dashboard specifications describe metrics that would be valuable but may not reflect what your actual data infrastructure can produce. Before building from this specification, validate each metric against the data sources available and the reporting capabilities of your specific tools.*

E-04 Learning-to-Performance Gap Analysis *→ Chapter 9*

Designed to surface: A structured analysis of where Level 2 achievement fails to produce Level 3 transfer — with root cause classification of each gap by barrier type.

Prompt

I have Level 2 and Level 3 Enhancement data for [PROGRAM TITLE] and need to analyze the gap between training achievement and job performance.
Level 2 data (assessment outcomes by objective): [PASTE DATA]
Level 3 data (transfer survey / observation data): [PASTE DATA]
Job context where transfer should be occurring: [DESCRIBE]
Known transfer barriers: [DESCRIBE IF AVAILABLE]

Please analyze:
1. Which objectives show strong Level 2 but weak Level 3 transfer?
2. Which objectives show weak Level 2 AND weak Level 3?
3. For Category 1 (learned but not transferred), classify the likely barrier:
   - Transfer-inappropriate practice (training conditions too different from job)
   - Environmental barriers (tools, time, support missing on the job)
   - Motivational barriers (no incentive or reinforcement for new behavior)
   - Retention failure (learning occurred but decayed before application)
4. For each barrier type, the specific design change most likely to address it

Flag any pattern suggesting the objectives may need revision rather than the instructional design serving them.

Known limitations: *Transfer barrier classification requires organizational knowledge that survey data alone cannot provide. A 'motivational barrier' looks identical in the data to a 'management culture barrier' — both produce non-transfer, but one requires a design change and the other requires a leadership intervention. Supplement this analysis with direct conversations with managers and learners before acting on the classification.*

CROSS-CUTTING *Leadership, Management, and Solo Practice*

XC-01 Portfolio Health Audit *→ Chapter 13*

Designed to surface: A cross-portfolio analysis of strategic alignment, coverage gaps, quality evidence, currency, and capacity fit — the structured visibility into a portfolio that typical program-by-program management does not provide.

Prompt

```
I need to conduct a health audit of my L&D program portfolio.

 Portfolio inventory: [PROGRAM NAMES, DATES LAST REVIEWED,
EVALUATION DATA,
   ORGANIZATIONAL PRIORITY SERVED]
 Current strategic priorities: [LIST 3-5]
 Team capacity: [FTE AVAILABLE FOR MAINTENANCE AND DEVELOPMENT]
 Most recent leadership feedback on L&D priorities: [DESCRIBE]

Please conduct a portfolio health analysis across five dimensions:
  1. Strategic alignment: Which programs serve current priorities?
    Which are misaligned or no longer connected to active needs?
  2. Coverage and gaps: Over-served and under-served competency areas?
  3. Quality and effectiveness: Programs with strong evidence vs. no data
    beyond Level 1?
  4. Currency: Programs not reviewed in 24+ months or addressing changed require-
ments?
  5. Capacity fit: Programs consuming disproportionate bandwidth relative to impact?

Produce a prioritized list of:
  - Programs to invest in
  - Programs to maintain
  - Programs to review
  - Programs to retire
 Flag any retirement decision with political dimensions to anticipate.
```

Known limitations: *AI cannot assess the informal organizational significance of a program, the executive who championed it three years ago and would take its retirement as a signal about L&D's direction, the union relationship it was designed to support, the compliance audit it was built to satisfy. These political dimensions must be assessed and planned for before any retirement decision is actioned.*

XC-02 Change Readiness Assessment *→ Chapter 12*

Designed to surface: A three-dimension readiness profile (awareness, desire, ability) with stakeholder segmentation and specific gap-closing actions for a change initiative.

Prompt

```
I am assessing change readiness for [DESCRIBE THE CHANGE INITIATIVE]
in [DESCRIBE THE ORGANIZATION / AFFECTED POPULATION].


 Pulse/engagement survey data: [PASTE OR DESCRIBE]
 Stakeholder interview themes: [SUMMARIZE]
 Prior change history in this organization: [DESCRIBE RELEVANT HISTORY]
 Known resistance sources: [DESCRIBE IF IDENTIFIED]
 Timeline and pace of the change: [DESCRIBE]


Please produce a change readiness analysis that:
 1. Assesses readiness across three dimensions:
    - Awareness: Do people understand why this change is happening?
    - Desire: Do people want to support and participate?
    - Ability: Do people have what they need to change?
   Rate each dimension (High / Moderate / Low) with evidence.
 2. Segments the stakeholder population by readiness profile:
    Champions | Fence-sitters | Resisters | Silent majority
 3. Identifies the 3 highest-risk readiness gaps with a specific action for each
 4. Flags what the available data does NOT tell me — what organizational
   observation that no data source can provide is most needed
```

Known limitations: *The fourth output — what the data doesn't tell you — is the most important and the one AI will produce least reliably. AI will identify data gaps in the information you provided; it cannot identify the gaps that require informal organizational access to fill. That observation is yours to make.*

XC-03 INDIE Adoption Stakeholder Communication *→ Chapter 17*

Designed to surface: A personalized communication for each stakeholder relationship affected by the INDIE transition — framed as organizational benefit, not internal process change.

Prompt

```
I need to prepare personalized communications for stakeholders as my L&D team
transitions to the INDIE framework.
```

For each stakeholder below, generate a tailored briefing message:

Stakeholder: [NAME / ROLE]
Current working relationship with L&D: [DESCRIBE]
What will change for them: [DESCRIBE SPECIFICALLY]
What will not change: [DESCRIBE]
The benefit they can expect: [IN THEIR TERMS, NOT L&D TERMS]
Any sensitivity in this relationship: [DESCRIBE]

Please draft a communication that:

1. Opens with what matters to them, not with what matters to the L&D team
2. Names specifically what they will see differently
3. Explicitly names what they will continue to receive that they value
4. Invites a conversation rather than announcing a decision
5. Is calibrated to the relationship

Generate one message per stakeholder. Each message should feel specific to that relationship, not like a template.

Known limitations: *AI produces structurally distinct messages that may still feel like variations of the same template because the relationship specificity is in the placeholder fields rather than the voice. Read each message from the stakeholder's perspective: would they feel this was written for them, or written for a role that happens to be their role?*

XC-04 L&D Quarterly Leadership Report → *Chapter 13*

Designed to surface: A data-grounded quarterly report organized by strategic priority served — not program names — that presents evidence honestly, acknowledges gaps, and makes a specific request.

Prompt

I need to prepare a quarterly L&D report for [CHRO / EXECUTIVE LEADERSHIP].

Portfolio overview: [NUMBER] active programs, [NUMBER] in development
Team size: [NUMBER] FTE
Key data available: [LIST DATA SOURCES]
Most significant accomplishments this quarter: [LIST 3-4]
Most significant challenges: [LIST 2-3]
Portfolio decisions made: [DESCRIBE]
Strategic priorities the function served: [LIST]

Please draft a quarterly report that:

1. Opens with a one-paragraph executive summary in business terms
2. Presents portfolio data organized by strategic priority served (not by program name)
3. Shows the strongest evidence of impact at Level 3 or 4 for programs

that have it

4. Honestly names programs without strong evidence and the plan to address them
5. Identifies one area where the function needs investment or support, with a specific request
6. Closes with the priority focus for next quarter

Audience: Financially literate executives who value directness and evidence over polished summaries that avoid difficult information.

Known limitations: *AI quarterly reports default to positive framing, leading with wins, softening gaps. The most credible version of this report leads with the full picture, including programs without evidence. Review specifically for whether the honest account of gaps is present and legible, not buried in qualifications.*

XC-05 The Missing Colleague Review → *Chapter 15*

Designed to surface: A genuinely critical perspective on a deliverable from the viewpoint of a skeptical first-time reader — designed to catch what self-review misses.

Prompt

I need a critical review of the following [DELIVERABLE TYPE] before it goes to [CLIENT / STAKEHOLDER / PRODUCTION].

[PASTE CONTENT TO REVIEW]

Context:
This is for: [DESCRIBE CLIENT AND THEIR CONTEXT]
Audience for this deliverable: [WHO WILL READ / ACT ON THIS]
What I'm most uncertain about: [NAME YOUR OWN DOUBT IF YOU HAVE ONE]

Please read this as a skeptical, intelligent [CLIENT / COLLEAGUE / LEARNER] encountering it for the first time. Do not read it charitably.

Tell me:

1. What assumption did I make that I didn't state?
2. What is the weakest part of this, and what specifically makes it weak?
3. What question would this raise for the intended audience that I haven't answered?
4. What would you change if this were yours?
5. Is there anything that is technically correct but would not land well with this specific audience or in this specific context?

Do not tell me what is good about it. Tell me what to fix.

Known limitations: *This prompt's effectiveness depends entirely on the instruction 'do not read it charitably.' Without that instruction, AI defaults to balanced feedback. The limitation is that AI cannot fully adopt the specific skepticism of a person with context-specific knowledge: the client who has been burned by a similar approach, the colleague who knows the organizational history. Use this as a first pass, not a final check.*

GOVERNANCE *Technology, Ethics, and Quality Standards*

GOV-01 Technical Constraint Definition for IN Phase → *Chapter 14*

Designed to surface: A structured interview guide for surfacing LMS capabilities, data architecture requirements, and technical constraints before they become late-stage design problems.

Prompt

```
I am an L&D designer beginning the IN phase for [PROGRAM TITLE].
I need to define technical constraints before finalizing curriculum architecture.

  Target audience and organizational context: [DESCRIBE]
  Anticipated delivery formats: [eLearning, ILT, blended, mobile, etc.]
  Enhancement data requirements: [DESCRIBE WHAT WE NEED TO MEASURE]
  Timeline: [DESCRIBE]

Please generate a structured 45-minute interview guide for the learning
technologist covering:
  1. LMS capabilities and limitations relevant to this program
     (completion tracking, mobile support, xAPI vs. SCORM, reporting granularity)
  2. Authoring tool constraints
     (interaction types, file size, browser/device compatibility, AI integration)
  3. Data architecture for Enhancement
     (what learner behavior data can be captured, at what latency, in what format)
  4. Integration requirements
     (HRIS connections, SSO, performance system linkage, dashboard tools)
  5. Known constraints from previous programs in this environment

Format as questions the designer asks, not a checklist the technologist fills out.
The goal is a conversation, not a form.
```

Known limitations: *AI generates technically appropriate questions, but the most valuable questions to ask your learning technologist are the ones specific to your organization's known infrastructure problems — the LMS configuration issue that has produced inaccurate data for the last two programs, the authoring tool version that doesn't support the interaction type you planned to use. Add those before the conversation.*

GOV-02 AI Tool Evaluation Framework → *Chapter 14*

Designed to surface: A technical assessment framework covering integration compatibility, data privacy, output auditability, vendor stability, and total cost of integration — for any AI tool under consideration.

Prompt

My L&D team is evaluating AI tools to integrate into our workflow.
I am the learning technologist responsible for the technical assessment.

Tools under consideration: [LIST TOOLS AND INTENDED USE]
Existing infrastructure: [LMS, authoring platform, analytics tools]
Organizational AI governance requirements: [DESCRIBE IF ANY]
Data privacy context: [INDUSTRY, LEARNER DATA SENSITIVITY, RELEVANT REGULATIONS: GDPR, FERPA, HIPAA, ETC.]

Please produce a technical evaluation framework for each tool covering:
1. Integration compatibility: Does it connect to existing infrastructure?
2. Data handling and privacy: What learner data does it process? Where is it stored? What are the retention and deletion policies?
3. Output quality and auditability: Can outputs be traced to source data? Is there a human review step built into the tool's design?
4. Vendor stability and support: How established is the vendor? What happens to our data if the vendor changes or closes?
5. Total cost of integration: Not just licensing — implementation time, training, ongoing maintenance, and switching cost

Produce a scoring rubric for each dimension and a recommendation format.

Known limitations: *AI evaluation frameworks assess what is generally relevant across tools; they cannot assess the specific data handling practices of the specific vendor you are evaluating, which change frequently and require direct review of the vendor's current privacy policy, DPA, and subprocessor list. Do not substitute this framework for that review.*

GOV-03 SME Pre-Review Briefing for AI-Generated Content → *Chapter 14*

Designed to surface: A briefing that directs SME attention toward the specific failure modes of AI-generated content in their domain — preventing the surface-level review that misses technical errors.

Prompt

I need to brief a subject matter expert before they review AI-generated content for [PROGRAM TITLE].

SME role: [DESCRIBE]
Content to be reviewed: [DESCRIBE — scenarios, procedural content, etc.]
The specific AI failure modes most likely for this content type: [DESIGNER FILLS IN — e.g., for clinical content: values on depicted equipment, protocol currency, anatomical accuracy]

Please draft a pre-review briefing that:
1. Explains in plain language how the content was produced (AI-generated draft, reviewed by designer, now needs domain accuracy review)

2. Tells them specifically what to look for, given the likely failure modes
   for this content type
3. Tells them what NOT to spend time on:
   - Writing style and formatting (the designer's domain)
   - Whether scenarios reflect best practice as they understand it
     (flag what IS current practice first)
   - Complete rewrites (flag the issue, don't fix it)
4. Gives them a simple annotation format:
   FLAG: [element] — [what's wrong] — [what it should be]
5. Sets a realistic time expectation

Tone: Collegial, specific, respectful of their time.

Known limitations: *The most important element of this briefing — the specific failure modes for this content type — cannot be generated by AI because it requires knowing what AI typically gets wrong in your specific domain. You must fill in item 2 based on your own experience with AI-generated content in this area.*

A Note on Prompt Maintenance

AI models are updated regularly, and the same prompt may produce meaningfully different output as the underlying model changes. If a prompt that previously worked well begins producing noticeably different output, revisit the prompt before concluding that the approach is failing. What changes in AI model updates: reasoning depth, default tone, response structure, and the specific ways the model handles ambiguous instructions. What rarely changes: the fundamental value of specific context, the improvement that comes from assigning a perspective, and the necessity of expert review for outputs that inform design decisions.

Document prompt improvements in your team's shared library when you find them. The prompts in this appendix were effective at the time of writing. Your practice will make them better.

Appendix B
Theoretical Quick Reference

This appendix provides concise practitioner-facing summaries of the seven learning theories referenced throughout the book. Each entry is written for use, not for citation, the goal is a working understanding that informs design decisions, not academic completeness. For deeper engagement with any framework, the Further Reading list in Appendix E identifies primary sources and recommended commentaries.

Andragogy: Malcolm Knowles	
Theorist / Era	Malcolm Knowles (1913–1997), American adult educator. Developed and popularized the concept of andragogy as distinct from pedagogy through 'The Modern Practice of Adult Education' (1970) and subsequent works.
Original context	Mid-20th century adult education reform, reaction against child-focused pedagogical models being applied uncritically to adult learners.
Core insight	Adults learn differently from children in ways that require a different instructional approach. Adult learning is self-directed, experience-based, problem-centered, and motivated by immediate applicability rather than future use.
Key principles	• Adults are self-directing and resist instruction that treats them as passive recipients • Adults bring accumulated experience that is both a resource for learning and

	a source of resistance when it conflicts with new information • Adults' readiness to learn is tied to their social roles and life situations • Adults are problem-centered rather than subject-centered: they want to learn what helps them address real challenges now • Adults are internally motivated; external motivation (grades, compliance) is less effective than intrinsic relevance • Adults need to know why they are learning something before they engage seriously with it
In INDIE	Drives the IN-Phase emphasis on performance problems over content topics, the learner persona development focus on job challenges and WIIFM, and the objective development standard that connects learning to immediate job application. Informs Event 2 (learner-facing objective presentation) and Event 3 (prior experience activation).
Practitioner caution	Andragogy describes tendencies, not universal laws. Individual adults vary significantly in their self-directedness, comfort with prior knowledge being challenged, and preference for problem-centered vs. structured learning.

Social Constructivism: Lev Vygotsky

Theorist / Era	Lev Vygotsky (1896–1934), Soviet psychologist. Primary works translated posthumously; 'Thought and Language' (1962) and 'Mind in Society' (1978) introduced his ideas to Western audiences.
Original context	Early Soviet psychology, response to behaviorism and individually-focused cognitive psychology. Died at 37 before fully developing his framework.
Core insight	Learning is fundamentally social. Knowledge is constructed through interaction with more capable others and through culturally-mediated tools and language. The Zone of Proximal Development (ZPD), the space between what a learner can do independently and what they can do with support, is the productive learning zone.
Key principles	• The Zone of Proximal Development is where learning most efficiently occurs: neither too easy nor unachievably hard • More knowledgeable others (teachers, peers, experts) mediate learning through scaffolding, structured support that is gradually removed as competence develops • Language and cultural tools are not just communication media but cognitive tools that shape thinking

	• Social interaction precedes and drives individual cognitive development • Learning is situated in social and cultural context; abstract learning divorced from context transfers poorly
In INDIE	Drives the IN phase ZPD estimation in learner personas, the curriculum architecture scaffolding design (Principle 2: supported to independent performance), the graduated practice in Event 6 (D phase), and the community of practice emphasis in team development (Chapter 14). Also underpins the Chapter 16 argument that solo practitioners need professional community for genuine intellectual development.
Practitioner caution	The ZPD is a useful design target but difficult to identify precisely without direct observation of learners. Estimated ZPDs from personas should be treated as hypotheses to test in the pilot, not precision measurements. The social learning emphasis can be over-applied to contexts where individual quiet reflection is more appropriate than group discussion.

Conditions of Learning: Robert Gagné	
Theorist / Era	Robert Gagné (1916–2002), American educational psychologist. 'The Conditions of Learning' first published 1965, revised through four editions. Nine Events of Instruction elaborated across subsequent works.
Original context	Mid-20th century instructional science and military training research. Influenced by both behavioral and cognitive psychology traditions; attempted to synthesize them.
Core insight	Different types of learning (intellectual skills, cognitive strategies, verbal information, motor skills, attitudes) require different conditions and instructional approaches. The Nine Events of Instruction describe the psychological conditions that must be established for learning to occur, regardless of content domain.
Key principles	• Learning outcomes are categorized into five types, each requiring specific internal and external conditions • The Nine Events describe a sequence of instructional activities that systematically establish learning conditions • Event 1 (attention) is foundational: without it, none of the subsequent events function

	• Event 3 (stimulate recall) is the cognitive bridge between prior knowledge and new learning • Event 6 (elicit performance) is non-negotiable: learning that doesn't require doing produces retention, not capability • Event 9 (enhance retention and transfer) is designed for, not assumed to happen naturally
In INDIE	Provides the structural architecture for Chapter 7 (Phase D) and the Development QA checklist. Each module in INDIE is auditable against the nine events. The Gagné framework is the primary tool for distinguishing instruction (which creates learning conditions) from content delivery (which presents information).
Practitioner caution	The nine events are necessary conditions for learning, not a rigid script. Skilled facilitation often merges events or sequences them differently based on audience and context. The framework describes what must happen psychologically, not the exact choreography. Mechanical adherence to the event sequence can produce technically complete but experientially wooden instruction.

Cognitive Development and Schema Theory: Jean Piaget

Theorist / Era	Jean Piaget (1896–1980), Swiss developmental psychologist. Primarily studied child development; his schema theory has been applied extensively to adult learning. Key works include "The Psychology of Intelligence" (1950).
Original context	Early 20th century child psychology. Piaget's stage theory of child development is largely superseded, but his schema theory of knowledge organization remains foundational to cognitive learning science.
Core insight	Knowledge is organized in cognitive schemas, mental frameworks that represent patterns of understanding. Learning occurs through assimilation (fitting new information into existing schemas) or accommodation (restructuring existing schemas to incorporate incompatible information). Prior knowledge is not just context for new learning; it actively shapes what can be learned.
Key principles	• Schemas are the cognitive structures through which we interpret experience and organize knowledge • Assimilation adds new information to existing schemas without changing their structure • Accommodation restructures existing schemas when new information cannot be assimilated

	• Equilibration is the driving force: cognitive dissonance between new information and existing schemas motivates restructuring • Prior knowledge is simultaneously a resource (existing schemas facilitate assimilation) and a constraint (deeply held schemas resist accommodation)
In INDIE	Drives the IN-Phase needs analysis classification of gap types (missing schemas vs. outdated schemas vs. schema extension), informing whether instruction should build forward or address prior misconceptions first. Also informs Event 3 design: effective recall activation surfaces schemas specifically to connect them to new content, or to surface misconceptions before they resist accommodation.
Practitioner caution	The schema framework is useful for diagnosing gap types but is not a precision instrument, determining whether a gap reflects a missing or outdated schema requires direct investigation of what learners currently know, not inference from performance data alone. Piaget's stage theory of cognitive development does not apply to adult learners and should not be used to characterize adult learning capacity.

Performance Objectives: Robert Mager	
Theorist / Era	Robert Mager (1923–2020), American behavioral psychologist and instructional technologist. 'Preparing Instructional Objectives' (1962) established the three-part objective format that remains the field standard.
Original context	Mid-20th century programmed instruction and behavioral psychology. Mager sought to make instructional intent explicit and measurable, moving away from vague educational goals.
Core insight	The purpose of instruction is to change behavior in observable, measurable ways. Learning objectives must specify exactly what learners will be able to do, under what conditions, and to what standard, making instructional intent testable rather than aspirational.
Key principles	• Every learning objective must specify Performance: what the learner will DO (observable behavior, active verb) • Every learning objective must specify Conditions: the circumstances under which the performance will occur • Every learning objective must specify Criterion: the standard against which performance will be judged

	• Objectives that cannot be assessed are not objectives, they are intentions • The objective is the primary quality criterion for assessment design: assessments that don't measure the stated objective are not valid • Objectives should reflect job performance requirements, not content coverage

In INDIE	The mandatory standard for all objectives produced in the IN phase. The IN-07 and IN-08 prompts are built around Mager's three-part format. The Development QA checklist includes Mager compliance as a required criterion. The Enhancement phase connection: if objectives are not Mager-compliant, Level 2 assessment validity cannot be established.
Practitioner caution	Mager's format describes necessary conditions for a good objective but not sufficient ones. An objective can be fully Mager-compliant (Performance + Condition + Criterion all present) and still address the wrong gap, require the wrong cognitive level, or set a criterion that is not achievable in the available learning time. Mager compliance is a format check, not a quality guarantee.

Four Levels of Evaluation: Donald Kirkpatrick	
Theorist / Era	Donald Kirkpatrick (1924–2014), American training and development professional. Published the four-level model in a series of journal articles (1959–1960); systematized in 'Evaluating Training Programs' (1994). Extended by son James Kirkpatrick.
Original context	Mid-20th century training evaluation, reaction against the absence of systematic measurement in organizational training. Provided a common framework for a field that had little shared evaluation language.
Core insight	Training effectiveness must be evaluated at multiple levels, each providing different information about different outcomes. Level 1 (reaction) measures how learners felt about training. Level 2 (learning) measures whether learning occurred. Level 3 (behavior) measures whether transfer to job performance occurred. Level 4 (results) measures organizational impact.
Key principles	• Level 1 (Reaction): Learner satisfaction and perceived relevance. Quick to collect; weak predictor of learning • Level 2 (Learning): Knowledge, skill, and attitude change. Direct measure of instructional effectiveness • Level 3 (Behavior): Transfer of learning to job performance. Most consequential for organizational impact; hardest to measure

	• Level 4 (Results): Organizational outcomes attributable to training. Most important to leadership; requires predefined metrics and baseline • Higher levels require more resources and time but provide more meaningful evidence of program value • Level 4 measurement requires Level 4 thinking in program design: metrics must be defined before the program launches
In INDIE	Provides the evaluative architecture for Phase E (Enhancement). Each Kirkpatrick level becomes a continuous data stream in INDIE rather than a periodic measurement event. The Enhancement dashboard is designed around all four levels. The tuning vs. replacing decision table in Chapter 9 maps data patterns from each level to specific responses.
Practitioner caution	Kirkpatrick correlation is not causation: strong Level 4 results following a training program do not establish that the training caused the results. Present Level 4 findings with appropriate methodological caveats. Also: high Level 1 scores are not evidence of learning; many well-liked programs produce poor Level 2 and 3 outcomes. Do not use Level 1 data as a proxy for program quality.

Taxonomy of Educational Objectives: Benjamin Bloom	
Theorist / Era	Benjamin Bloom (1913–1999), American educational psychologist. Led the committee that produced 'Taxonomy of Educational Objectives: Handbook I, Cognitive Domain' (1956). Revised taxonomy published by Anderson and Krathwohl (2001).
Original context	Mid-20th century educational psychology and curriculum reform. Sought to provide a common language for describing educational objectives across levels and disciplines.
Core insight	Cognitive learning objectives exist on a hierarchical continuum from lower-order thinking (recall and recognition) to higher-order thinking (analysis, synthesis, evaluation). The cognitive level of an objective must match the cognitive demand of the performance it is meant to develop.
Key principles	• Knowledge (Remember): Recall of facts, definitions, procedures without understanding • Comprehension (Understand): Interpretation, explanation, paraphrasing of information • Application (Apply): Using learned information to solve problems in new situations

	• Analysis (Analyze): Breaking information into components; identifying patterns and relationships • Synthesis (Evaluate in revised taxonomy): Judging value, making decisions, defending positions • Evaluation (Create in revised taxonomy): Producing new work, designing solutions, constructing original responses • Most organizational training targets Application or Analysis; objectives written at Knowledge level for Application-level jobs represent a systematic design error
In INDIE	Drives the Bloom's level requirement in IN-07 and IN-08 (objective development and quality review), the cognitive level specification in D-06 (practice design), and the assessment item development in D-08. The Development QA checklist includes Bloom's alignment as a required criterion. The rule: the cognitive level of practice and assessment must match the Bloom's level of the objective.
Practitioner caution	Bloom's levels are not a strict hierarchy in practice: real performance often requires simultaneous operation at multiple levels. The taxonomy is most useful as a diagnostic tool (is this objective at the right level for what the job requires?) rather than as a rigid sequencing framework. The 2001 revised taxon-

	omy changed the naming conventions; practitioners should verify which version their organization or partners use to avoid miscommunication.

Additional Frameworks Referenced in the Book

Several additional frameworks are referenced in the book but not included above because they are role-specific rather than universally applicable:

- **Gilbert's Behavior Engineering Model (Chapter 13):** The performance consulting root cause framework distinguishing environmental from person factors. Essential for performance consultants and OD specialists; informative for all designers approaching the IN phase.
- **Sweller's Cognitive Load Theory (Chapter 7):** Intrinsic, extraneous, and germane load. The essential complement to Gagné in the Development phase; drives the five-concept maximum and the cognitive load audit standard in the Development QA checklist.
- **Lave and Wenger's Situated Learning and Communities of Practice (Chapters 12, 16):** The argument that learning is embedded in practice communities rather than occurring in isolated cognition. Drives the community of practice emphasis in team development and the professional isolation concern for solo practitioners.
- **Rummler and Brache's Performance System Framework (Chapter 13):** The three-level view of organizational performance (organization, process, job). Positions performance consulting within a systems thinking tradition and explains why most training requests are actually process or organization-level problems.

Appendix C
Role Comparison Matrix

This matrix maps the eleven roles covered in Part Three against each INDIE phase, describing the primary contribution each role makes at each stage. Use it to navigate to the most relevant chapter sections for your current role and project phase, and to plan role integration when multiple practitioners are working on the same program.

Note: In many organizations, one practitioner holds multiple roles simultaneously. Where roles overlap, the matrix shows the full range of contributions that the combined role must make, which is the most honest account of what solo or small-team practice actually requires.

Role	PHASE IN Investigation-Design	PHASE D Development	PHASE I Implementation	PHASE E Enhancement
Instructional Designer	Leads: stakeholder analysis, objective writing, curriculum architecture. Owns the performance consulting judgment about whether training is appropriate.	Drives Gagné event coverage, cognitive load audit, assessment design. AI review authority for pedagogical quality.	Supports facilitation guide, scope management, pilot observation. Provides design expertise for rapid revisions.	Analyzes Level 2 data by objective, learning-to-performance gap analysis, program revision decisions.
Learning Experience Designer	Leads: learner persona emotional profiling, empathy research synthesis, ZPD estimation. Co-designs curriculum with ISD.	Leads: emotional arc design, UX architecture, accessibility-first design. Reviews all scenarios for representation.	Reviews participant communications for tone and learner-centeredness. Monitors learner experience during pilot.	Analyzes Level 1 qualitative data for experience patterns. Leads UX-informed program revision.
eLearning Developer	Advises on technical feasibility of pro-	Builds all eLearning deliverables. Man-	Resolves technical issues during deployment. Supports	Monitors technical performance data. Implements

	posed interaction designs. Flags constraints early.	ages AI tool integration, rapid prototyping, and media asset quality.	facilitators on technology use.	content and interaction revisions based on Enhancement findings.
Multimedia Producer	Advises on media production constraints and timelines that affect design decisions.	Produces all multimedia assets. Makes AI voice decisions. Leads domain accuracy review of AI-generated visuals.	Resolves media delivery issues. Supports facilitators on media-dependent session elements.	Revises media assets based on learner feedback and accessibility audit findings.
Facilitator / Trainer	Provides ground-truth learner knowledge from prior delivery experience. Participates in integration meeting.	Reviews facilitation guide for practicability. Tests activity timing. Flags guide gaps before delivery.	Central role: delivers program, manages group dynamics, pilots and iterates. Provides debrief data for enhancement.	Primary source of Level 1 qualitative insight and facilitator debrief data. Informs design improvements.
Performance Consultant / OD Specialist	May lead entire IN phase: performance gap diagnosis, root cause analysis, intervention selection. Determines whether training is appropriate.	Advises on whether content reflects accurate performance requirements. Reviews for organizational authenticity.	May design change management alongside program deployment. Monitors organizational readiness.	Leads Level 3 and 4 analysis. Connects program outcomes to organizational performance system.
L&D Manager / Curriculum Coordinator	Sets portfolio context and strategic alignment. Approves scope. Manages stakeholder relationships that underpin the project.	Monitors quality standards across projects. Maintains shared prompt library. Approves deliverables before client review.	Manages rollout coordination, stakeholder communications, and the organizational conditions for successful implementation.	Owns portfolio health monitoring, quarterly leadership reporting, and Enhancement cadence maintenance across all programs.
Learning Technologist	Defines technical constraints before design decisions are finalized. Speci-	Provides technical review at Development milestones. Validates LMS configuration and	Manages technical deployment. Validates completion tracking before	Custodian of Enhancement data infrastructure. Ensures data quality for AI analysis

	fies data architecture for Enhancement.	SCORM/xAPI compliance.	full rollout. Resolves LMS issues during delivery.	tools. Maintains dashboard.
Subject Matter Expert	Provides domain knowledge for needs analysis. Participates in integration meeting. Primary source of conditional and contextual knowledge.	Reviews AI-generated content for domain accuracy. Validates scenarios against real practice. Confirms procedure currency.	May co-facilitate technical content. Available for learner questions beyond guide scope.	Reviews content revisions for ongoing accuracy. Flags domain changes that require content updates.
Project Manager	Facilitates integration meeting. Establishes project infrastructure: charter, RACI, risk log, communications plan.	Manages development milestones and review cycles. Owns scope change assessment and timeline management.	Coordinates deployment logistics. Manages stakeholder communications and issue resolution during rollout.	Tracks Enhancement cadence commitments. Escalates when sustaining rhythm is threatened by operational pressure.
Solo Practitioner / Consultant	Performs all IN phase roles simultaneously. AI accelerates data synthesis and persona drafting. Expert judgment remains the quality standard.	Produces all Development deliverables with AI assistance. Missing colleague review prompt compensates for absence of peer review.	Manages all client communications, deployment coordination, and facilitation preparation. AI handles drafting; judgment shapes every output.	Runs Enhancement with AI synthesis support. Practice health check prompt maintains strategic alignment without managerial oversight.

Reading the Matrix

- **Bold text in the Role column** indicates that this role is the primary owner of that phase dimension. Where text is not bold, the role is a contributor or reviewer rather than the lead.
- **"Leads"** indicates the role has both the primary analytical or production responsibility and the decision authority for that work.
- **"Advises"** indicates the role provides expert input that shapes decisions owned by another role.
- **"Supports"** indicates the role contributes to work led by others, typically in a review, coordination, or technical capacity.

For roles not listed here (regional training coordinator, HR business partner, executive sponsor, etc.), use the matrix as a diagnostic: identify which listed role most closely approximates the responsibilities of the unlisted role, then use that row as the starting point for thinking about INDIE application.

Appendix D
AI Tool Landscape and Evaluation Rubric

This appendix provides two things: a category-organized overview of the AI tool types relevant to instructional design practice, and a weighted evaluation rubric for comparing specific tools within any category. It does not name or endorse specific products, because the AI tool landscape changes faster than any book can accommodate. The categories are stable; the specific products within them are not.

Use the category overview to understand what to look for when evaluating tools in each functional area. Use the rubric when comparing two or more tools for adoption. The rubric produces a weighted score between 1 and 4; it is a decision-support tool, not a decision-maker — context and organizational fit still require human judgment.

Tool Categories

Large Language Models (LLMs) — General Purpose
The foundational AI writing and analysis tools: large language models accessed directly through a chat interface or API. These are the tools behind every prompting workflow in Appendix A. They draft, synthesize, analyze, generate, and respond to complex instructions. In instructional design, they are the primary tool for content drafting, scenario generation, objective writing, feedback scripting, and analysis tasks across all four INDIE phases. **Tool types in this category (not product endorsements):** Major provider chat interfaces (accessed by subscription or free tier); API access for custom workflow integration; integrated versions embedded in productivity tools **Watch for:** Context window limits affect quality on long documents; knowledge cutoff dates produce outdated information in rapidly changing domains; outputs present false information with the same fluency as correct information. The most important evaluation criterion: does this tool perform well on the specific tasks you need it for, not on benchmark tasks?

AI Image Generation

Tools that generate visual assets from text descriptions. In instructional design, used for scenario character images, background environments, icon sets, simple diagrams, and custom illustrations that replace or supplement stock photography. Significantly faster and less expensive than custom illustration or photography for many use cases.

Tool types in this category (not product endorsements): Standalone image generation platforms; image generation integrated into design tools; API-based generation for programmatic workflows

Watch for: The five failure modes from Chapter 11: embedded text errors, anatomical inaccuracies, equipment configuration errors, and cultural context errors. Every image depicting specialized equipment, processes, or documents requires domain accuracy review by a qualified SME.

AI Voice Synthesis

Tools that convert text to spoken narration using AI-generated voices. In instructional design, used for procedural narration, high-volume content requiring frequent updates, and multilingual delivery where recording multiple human narrators is cost-prohibitive. Not appropriate for content requiring emotional authenticity, leadership voice, or human-to-human signal.

Tool types in this category (not product endorsements): Dedicated voice synthesis platforms with multiple voice options; voice cloning tools (subject to significant ethical constraints); narration features integrated into eLearning authoring tools

Watch for: The three AI voice failure modes from Chapter 11: pacing (absence of breath pauses), emphasis (default to linguistic patterns rather than instructional intent), and emotional register (trained to sound professional and pleasant, not grave or urgent). Always review synthesized audio at the paragraph level before approving for production.

AI-Assisted Authoring Tools

eLearning authoring platforms that have integrated AI capabilities directly into their interfaces. These tools may offer AI-generated quiz questions from content, suggested interactions, AI-written script alternatives, or AI-powered accessibility checking. The integration of AI into authoring tools is evolving rapidly; capabilities available at time of reading may differ significantly from what was available at time of writing.

Tool types in this category (not product endorsements): Major eLearning authoring platforms with integrated AI features; AI-augmented rapid development platforms; AI-powered course generation tools that build modules from documents or slides

Watch for: AI features integrated into authoring tools inherit all the quality limitations of the underlying AI models, with the additional risk that they are less visible than a separate AI interface. Treat AI-generated content in authoring tools with the same review standards as content generated in any other AI interface. AI-generated quiz questions are particularly likely to require distractor quality review.

AI Analytics and Enhancement Tools

Platforms that apply AI to learning analytics data: synthesizing open-ended survey responses at scale, identifying patterns in assessment performance data, predicting at-risk learners, generating natural language summaries of learning data for different audience dashboards. These tools are the operational infrastructure for INDIE's Enhancement phase at scale.

Tool types in this category (not product endorsements): LMS-integrated analytics platforms with AI reporting features; standalone learning analytics tools; survey platforms with AI text analysis; custom analytics built on LLM APIs

Watch for: AI analytics tools are only as reliable as the data they process. Poor data architecture (malformed xAPI data, inaccurate completion tracking, survey response rates below 30%) produces analytics that appear authoritative but are based on insufficient or unreliable data. The learning technologist's data quality role (Chapter 15) is the prerequisite for these tools to function meaningfully.

AI Transcription and Captioning

Tools that convert spoken audio and video to text, enabling automated captioning, interview transcript generation, and accessibility compliance workflows. In instructional design, used for stakeholder interview transcription (supporting IN phase synthesis), facilitator debrief notes, and video caption generation. One of the highest-value AI tools for instructional design in terms of time saved relative to implementation cost.

Tool types in this category (not product endorsements): Standalone transcription platforms; video platform integrated captioning; transcription integrated into recording and editing tools; open-source transcription models

Watch for: Auto-generated captions require word-for-word review before publication. Technical vocabulary, proper nouns, domain-specific terminology, and non-native speaker audio are the most common transcription failure points. The Chapter 11 rule applies unconditionally: never publish auto-generated captions without correction.

AI Video Generation and Editing

Tools that generate or significantly manipulate video content using AI. Capabilities include generating talking-head video from text, creating animation from scripts, removing backgrounds and objects, and assembling footage based on instructions. This category is evolving fastest and is most subject to quality variation between tools.

Tool types in this category (not product endorsements): AI video generation platforms; AI-enhanced screen recording and editing tools; avatar-based video tools; AI-powered video production assistants

Watch for: Quality thresholds for AI video vary widely and are improving rapidly.

Evaluation Rubric

Use this rubric when comparing specific tools within any category. Score each tool on each criterion using the 1–4 scale. Multiply each score by the criterion weight. Sum the weighted scores for a total between 1.0 and 4.0. Compare total scores across tools, but treat the score as one input to the decision rather than the decision itself.

Criterion	Wt.	What each score level means (1–4)
Output quality for intended use Does the tool produce outputs that reliably meet the quality standard required for this specific task in this specific context? Quality is task-specific: the correct question is not 'is this a good AI tool?' but 'does this tool produce good outputs for the task I need it for?'	25%	**1** — Outputs consistently require substantial revision or cannot be used **2** — Outputs usable after moderate revision; quality inconsistent **3** — Outputs usable after standard review; quality generally reliable **4** — Outputs consistently meet standards with minimal review required
Human review support Does the tool's design support or undermine the human review that responsible AI use requires? Some tools surface their confidence level or cite sources; others present outputs as finished products with no review prompts. Tools that make review easy and obvious are governance-supportive.	20%	**1** — No review support; outputs presented as finished products **2** — Minimal review support; limited transparency about limitations **3** — Adequate review support; limitations acknowledged in some contexts **4** — Strong review support; sources cited, limitations transparent, review steps built in
Data privacy and handling What learner or organizational data does this tool process, where is it	20%	**1** — Significant data privacy concerns; unclear policies; data used for model training **2** — Some privacy concerns; policies exist but complex or unfavorable

stored, and what are the retention policies? This criterion applies most stringently to tools that process individual learner data (assessment responses, behavioral analytics, written submissions). It applies with lower stringency to tools that process only designer-entered content.		**3** — Adequate privacy posture; clear policies; data not used for model training **4** — Strong privacy posture; clear policies, DPA available, regional compliance documented
Integration with existing workflow How much friction does this tool add to or remove from the existing workflow? Integration cost includes not just technical integration but the learning curve, the workflow interruption during adoption, and the ongoing cost of maintaining the integration as tools update.	**15%**	**1** — High friction; significant workflow disruption; poor integration with existing tools **2** — Moderate friction; workflow disruption during adoption; limited integration **3** — Low friction; reasonable adoption curve; adequate integration with existing tools **4** — Minimal friction; smooth adoption; strong integration with existing infrastructure
Vendor stability and support AI tool vendors vary enormously in stability, from established technology companies to well-funded startups to early-stage products that may not survive a funding cycle. The cost of a vendor failure is not just losing the tool — it is losing the data, the workflows, and the	**10%**	**1** — High stability concern; early-stage or under-funded; limited support **2** — Moderate stability concern; some track record; basic support **3** — Adequate stability; established presence; reasonable support **4** — Strong stability; established vendor; clear data portability; strong support

organizational knowledge built around it.		
Total cost of integration The full cost includes not just licensing but implementation time, training time for team adoption, ongoing maintenance (including updates as the tool changes), and switching cost if the tool proves inadequate. Low-cost tools with high implementation friction are frequently more expensive than higher-cost tools with low friction.	10%	**1** — High total cost relative to value; switching cost prohibitive **2** — Moderate total cost; value proposition unclear **3** — Reasonable total cost; value proposition clear; manageable switching cost **4** — Strong value proposition; total cost well within benefit; low switching cost

How to Use the Rubric Score

- **3.3 or above:** Strong candidate for adoption. Proceed to pilot with governance review.
- **2.7 to 3.3:** Viable candidate with identified concerns. Address the lowest-scoring criterion explicitly before full adoption.
- **2.0 to 2.7:** Marginal candidate. Consider whether a higher-scoring alternative exists, or whether the specific use case justifies accepting the limitations.
- **Below 2.0:** Significant concerns. Do not adopt without substantial mitigation for the failing criteria, particularly if data privacy (criterion 3) is below 2.

The rubric does not replace the data privacy due diligence described in Chapter 17. Even a tool scoring 4.0 overall requires review of the vendor's current privacy policy, DPA availability, and subprocessor relationships before processing learner data.

Appendix E
Further Reading and Professional Resources

This list is curated for practitioners, not academics. The criterion for inclusion is whether a resource has been consistently recommended by experienced practitioners as having changed how they think or work. Primary sources for the theoretical frameworks in Appendix B are included for completeness; the other sections prioritize practical value over comprehensiveness.

Note on being current: L&D and AI resources have highly variable shelf lives. Resources listed here were selected for durability, either because they address enduring principles or because they represent the best available treatment of a topic at the time of writing. For current AI developments, professional community sources (listed at the end) will be timelier than any book.

Learning Theory Foundations

Knowles, M.S., Holton, E.F., & Swanson, R.A. The Adult Learner (9th ed.). Routledge.

The primary source for andragogy. Denser than you expect from a theory that is often summarized in a bullet list. Worth reading in full for the nuance that the summaries miss, particularly the sections on how andragogy varies across individual differences and cultural contexts.

Vygotsky, L.S. Mind in Society: The Development of Higher Psychological Processes. Harvard University Press, 1978.

The accessible Western introduction to Vygotsky. The sections on the Zone of Proximal Development and the role of tools and language in cognition are the most directly applicable to instructional design practice. Better read in company with a commentator like Wertsch for context.

Gagné, R.M. The Conditions of Learning and Theory of Instruction (4th ed.). Holt, Rinehart and Winston, 1985.

The primary source for the conditions of learning and the nine events. More technical than practitioners usually encounter it through secondary sources. The taxonomy of learning outcomes is particularly useful for understanding why different types of learning require different instructional approaches.

Mager, R.F. Preparing Instructional Objectives (3rd ed.). CEP Press, 1997.

Still the clearest treatment of behavioral objectives available. Short, direct, and deliberately conversational. Mager's programmed text format is unconventional but effective. Required reading for any practitioner who writes learning objectives.

Sweller, J., Ayres, P., & Kalyuga, S. Cognitive Load Theory. Springer, 2011.
The rigorous treatment of cognitive load theory for practitioners who want more than the three-type summary. The chapters on worked examples and the split-attention effect have the most direct implications for Development phase design decisions.

Brown, P.C., Roediger, H.L., & McDaniel, M.A. Make It Stick: The Science of Successful Learning. Harvard University Press, 2014.
The most accessible practitioner-facing treatment of cognitive science applied to learning design. The chapters on retrieval practice, spaced practice, and interleaving directly support the Event 9 design principles in the book. One of the most frequently recommended books in the field for good reason.

Instructional Design Practice

Dick, W., Carey, L., & Carey, J.O. The Systematic Design of Instruction (8th ed.). Pearson, 2014.
The most rigorous treatment of systematic instructional design available. More comprehensive than ADDIE, more analytically demanding than most practitioners engage with. Worth reading as a corrective to the "just use a template" approach to needs analysis.

Clark, R.C. & Mayer, R.E. e-Learning and the Science of Instruction (5th ed.). Wiley, 2023.
The most evidence-based treatment of multimedia learning principles in eLearning contexts. Directly applicable to Development phase decisions about visual design, audio, and interactive element use. Updated editions track current research; prioritize the most recent edition.

Van Merriënboer, J.J.G. & Kirschner, P.A. Ten Steps to Complex Learning (3rd ed.). Routledge, 2017.
The most sophisticated instructional design framework for complex skill development. More demanding to implement than ADDIE but produces better learning outcomes for complex, multi-step performance. Particularly valuable for practitioners designing technical or clinical training.

Gilbert, T.F. Human Competence: Engineering Worthy Performance. ISPI/Pfeiffer, 1978 (reprinted).
The primary source for the Behavior Engineering Model. More philosophical than its practical applications suggest. The opening chapters on the "potential for improving human performance" provide the conceptual foundation that makes the BEM coherent rather than just a diagnostic checklist.

Rummler, G.A. & Brache, A.P. Improving Performance: How to Manage the White Space on the Organization Chart (3rd ed.). Jossey-Bass, 2012.
The primary source for the performance system framework. The "white space" concept, the processes that fall between functions and that determine most organizational performance outcomes, is the essential systems thinking corrective to training-as-default problem-solving.

AI in Education and Learning

Cope, B. & Kalantzis, M. A Pedagogy of Multiliteracies: Learning by Design. Palgrave Macmillan, 2015.
Not specifically about AI, but provides the theoretical foundation for thinking about how digital and AI-mediated environments change the nature of learning and literacy. More relevant to the adaptive learning systems discussion in the Conclusion than to day-to-day practice.

Selwyn, N. Should Robots Replace Teachers? AI and the Future of Education. Polity Press, 2019.
A deliberately contrarian and intellectually rigorous examination of AI in education that resists both techno-optimism and techno-pessimism. Essential reading for practitioners who want to think clearly about the future direction described in the book's Conclusion.

Luckin, R. Machine Learning and Human Intelligence: The Future of Education for the 21st Century. UCL Institute of Education Press, 2018.
One of the most thoughtful treatments of what AI capabilities mean for learning design from a researcher who understands both the technology and the pedagogy. The framework for "intelligent support for learning" is more nuanced than most practitioner-facing AI-in-education writing.

Mollick, E. Co-Intelligence: Living and Working with AI. Portfolio/Penguin, 2024.
The most practically useful general treatment of working with AI available at time of writing. Mollick's framing of AI as a "co-intelligence" rather than a tool or a threat is the closest to the orientation this book takes. Particularly valuable for practitioners trying to explain AI's role to skeptical colleagues.

Evaluation and Measurement

Kirkpatrick, J.D. & Kirkpatrick, W.K. Kirkpatrick's Four Levels of Training Evaluation. ATD Press, 2016.

The authorized update of the Kirkpatrick model, co-authored by Donald Kirkpatrick's son and daughter-in-law. Addresses Level 4 measurement more practically than earlier versions and clarifies how to work backward from results to design. More useful than the original 1994 edition for practitioners building Enhancement cycles.

Brinkerhoff, R.O. The Success Case Method. Berrett-Koehler, 2003.

An alternative evaluation approach that identifies the conditions under which training produces exceptional transfer — more useful for understanding what makes transfer work than what proportion of learners transferred. Particularly relevant for Level 3 analysis when understanding transfer mechanisms matters more than measuring transfer rates.

Phillips, J.J. & Phillips, P.P. ROI Methodology Basics. ATD Press, 2015.

The most systematic treatment of Level 4 measurement methodology. More rigorous than most practitioners need for day-to-day Enhancement work, but essential for practitioners who need to make credible Level 4 claims to skeptical executives.

Change Management and Leadership

Kotter, J.P. Leading Change. Harvard Business Review Press, 2012.

The most widely used change management framework in organizational settings. The eight-step model provides useful structure for the INDIE adoption roadmap in Chapter 18. Particularly relevant for practitioners who need to make the organizational case for a framework transition.

Heath, C. & Heath, D. Switch: How to Change Things When Change Is Hard. Crown Business, 2010.

More practically useful than Kotter for understanding the psychology of individual behavior change during organizational transitions. The "rider, elephant, and path" framework is a useful diagnostic for understanding why INDIE adoption fails in the patterns described in Chapter 18.

Edmondson, A.C. The Fearless Organization: Creating Psychological Safety in the Workplace. Wiley, 2018.

The most rigorous treatment of psychological safety — the organizational condition that makes honest quality review, failure reporting, and genuine learning from mistakes possible. Directly relevant to building the team culture that INDIE's quality governance requires.

Professional Communities and Continuing Development

Association for Talent Development (ATD): td.org

The largest professional association for L&D practitioners globally. TD Magazine, the ATD blog, and ATD conference sessions are the most consistent source of practitioner-facing content on current tools, methods, and research. The ATD Handbook is the most comprehensive single reference for the field.

Learning and Performance Institute (LPI): thelpi.org

UK-based professional body with strong practitioner community and capability framework. Particularly valuable for practitioners in the UK and Europe seeking professional credentialing and peer community.

eLearning Guild / Learning Guild: learningguild.com

The professional community most focused on the technology dimensions of L&D practice. The Guild's research reports on specific tools and platforms are the most reliable practitioner-facing research available on the eLearning development side of the field.

International Society for Performance Improvement (ISPI): ispi.org

The professional home for performance consulting and human performance technology practitioners. Essential community for practitioners focused on the performance consulting and OD dimensions of the field rather than the instructional design production dimensions.

r/instructionaldesign (Reddit) and LinkedIn L&D communities

Acknowledge the informal community that is often more current than formal associations. The Reddit community in particular provides rapid practitioner-to-practitioner knowledge sharing on tool questions, career questions, and emerging practice issues. Signal-to-noise is lower than curated sources; the most-upvoted threads on specific topics are typically reliable.

On Staying Current

The most important professional development investment in an AI-augmented field is not reading more books. It is maintaining active connection with practitioners who are doing the work right now, in real organizations, with the tools that exist today. The professional community sources above are more current than any book. The colleagues in your own team and network are more current still.

This book is a framework and a foundation. The practice that grows from it will be built from your own project experience, your team's accumulated learning, and the ongoing conversation with a field that is changing faster than any publication cycle can fully capture.

Appendix F
INDIE Project Overview: From Context Architecture to Final Product

This appendix maps a complete INDIE project from the foundational context architecture work through every phase to a delivered, continuously improving program. Use it as a reference at project kickoff, as a navigation tool when returning to the book mid-project, and as a communication aid when explaining the INDIE process to stakeholders who need the overview without the detail.

FOUNDATION · Context Architecture and RAG Setup Build the project knowledge base · Prepare source documents · Configure AI workspace · Draft current-state summary								
IN INvestigation-Design	→	**D** Development	→	**I** Implementation	→	**E** Enhancement	→	**E+** Continuous Improvement
Investigate Design Concurrently		Build Against Gagné's Nine Events		Pilot Iterate Roll Out		Measure All Four K-levels		Tune or Replace Ongoing

FOUNDATION · Context Architecture and RAG Setup

Before any IN-Phase work begins, establish the AI workspace that will support every subsequent task. The quality of all AI-assisted work throughout the project depends directly on the quality of this foundation.

Steps

- Identify and gather source documents: SOPs, regulatory references, existing training content, job descriptions, and any organizational data relevant to the performance domain
- Verify each document is the current approved version from the document control system
- Apply descriptive filenames: DocumentType_Subject_Version_Date
- Create a new Project (Claude) or equivalent workspace and upload source documents
- Write a custom system prompt grounding the AI in the uploaded documents and instructing it to acknowledge uncertainty rather than fabricate

• Draft the project knowledge base current-state summary: what is known, what is open, what the next decision point is • Add stakeholder notes, performance data, and learner research to the knowledge base as the project progresses: the foundation is a living document, not a one-time setup **Reference** *Chapter 4: Working with AI: Prompt Engineering and Context Architecture · Document Preparation Checklist · FACTS Evaluation Framework*

IN-PHASE **Investigation-Design**	*Fuse performance analysis and learning design as concurrent streams informed by each other from the first stakeholder conversation.*
Key Steps	• Conduct stakeholder interviews; synthesize findings using AI (IN-01, IN-02) • Identify and classify the performance gap: missing knowledge, outdated schema, or motivational/environmental cause (Gilbert BEM) • Develop root cause hypotheses and test them against available data (IN-03, IN-04) • Build learner personas with ZPD estimation and job-challenge profiling (IN-05, IN-06) • Write learning objectives in Mager format: Performance + Condition + Criterion, at the correct Bloom's level (IN-07, IN-08) • Design curriculum architecture: sequence, scaffolding from supported to independent performance, time allocation (IN-09) • Conduct integration meeting: align stakeholders, confirm scope, establish governance
Quality Gate	✓ Performance gap is confirmed and root cause classified ✓ All objectives meet Mager criteria (Performance + Condition + Criterion present) ✓ Bloom's levels match job performance requirements ✓ Stakeholders have validated the gap statement and approved the curriculum map
AI Prompts **IN-01 through IN-11** Appendix A	**Primary Deliverables** Needs analysis report with validated root cause Learner personas with ZPD estimates Learning objectives (all Mager-compliant, Bloom's-leveled) Approved curriculum map and sequence

<table>
<tr><td>D-PHASE
Develop-
ment</td><td>Build learning experiences against Gagné's Nine Events of Instruction as structural architecture, complemented by Sweller's Cognitive Load Theory.</td></tr>
<tr><td>Key Steps</td><td><ul>
<li>Design each module against all Nine Events: attention, objective, recall, content, guidance, practice, feedback, assessment, transfer (D-01 through D-09)</li>
<li>Generate content drafts, scenarios, case studies, and examples using AI — apply FACTS evaluation (Faithful, Accurate, Complete, Targeted, Safe) to every output</li>
<li>Conduct cognitive load audit: no more than five concepts per module; remove extraneous load; build germane load through worked examples (D-10)</li>
<li>Request SME domain accuracy review; validate all procedure steps, regulatory references, and equipment details against current approved documents</li>
<li>Produce assessment bank: items at the correct Bloom's level, matched to stated objectives</li>
<li>Review all AI-generated media (images, voice, video, captions) against quality standards; apply Chapter 11 failure mode checklist</li>
<li>Complete Development QA checklist: Nine Events coverage, Mager compliance, Bloom's alignment, cognitive load, SME sign-off, accessibility</li>
</ul></td></tr>
<tr><td>Quality Gate</td><td>√ All Nine Gagné Events are present and functional in each module
√ Cognitive load audit passed: five-concept maximum observed
√ SME has reviewed and approved domain accuracy
√ Every AI-generated output has passed FACTS evaluation
√ Development QA checklist signed off before content enters final production</td></tr>
<tr><td>AI Prompts
D-01 through D-14
Appendix A</td><td>Primary Deliverables
Complete learning modules (eLearning, ILT, or blended)
Facilitator guides and participant materials
Assessment bank with items matched to objectives
Media assets reviewed and approved</td></tr>
</table>

<table>
<tr><td>I-PHASE
Implementa-
tion</td><td>Learn fast enough to improve the program before full rollout. The pilot is a learning event, not a dress rehearsal.</td></tr>
</table>

Key Steps	• Select pilot cohort (8–15 participants): representative of the full audience, includes skeptics, includes at least one person who represents the most challenging learner segment (I-01) • Prepare facilitators: not just content familiarity but scenario anticipation, timing, and contingency planning (I-02, I-03) • Deliver pilot with structured observation: track engagement signals, confusion points, activity timing, and participant language • Conduct facilitator debrief immediately after pilot; synthesize findings using AI (I-04, I-05) • Identify rapid-iteration changes: distinguish critical (fix before rollout) from developmental (address in Enhancement cycle) • Communicate with stakeholders using pilot-thinking framing: learning event, not a launch • Execute full rollout with monitoring; resolve issues using established escalation path
Quality Gate	✓ Pilot debrief completed and documented before rollout begins ✓ Critical design issues resolved; rollout-blocking problems have no open items ✓ Completion tracking validated in LMS before full rollout ✓ Stakeholder communications have maintained pilot-thinking framing throughout

AI Prompts **I-01 through I-09** Appendix A	**Primary Deliverables** Delivered and completed program Pilot debrief report with categorized findings Rollout record and completion tracking confirmation Rapid-iteration log (what changed between pilot and rollout, and why)

E-PHASE **Enhancement**	*Run continuous improvement from launch forward. Enhancement is not evaluation that happens at the end — it is a data-driven improvement cycle that begins at delivery and runs for as long as the program runs.*

Key Steps	• Level 1 — Reaction: Collect and synthesize learner satisfaction and relevance data after each cohort; identify qualitative patterns using AI (E-01) • Level 2 — Learning: Analyze assessment performance by objective; identify objectives with consistent failure rates; flag for instructional redesign (E-02) • Level 3 — Behavior: Gather job transfer data through observation, manager reports, or performance metrics; connect transfer patterns to program design decisions • Level 4 — Results: Measure organizational outcomes attributable to the program; compare against pre-defined baseline metrics established before launch • Apply tuning vs. replacing decision framework: tune (revise content, sequence, or practice) when the design is sound but delivery conditions failed; replace when root cause analysis reveals a design error • Update the project knowledge base with Enhancement data from each cycle; this becomes the foundation for the next iteration • Deliver quarterly leadership report connecting program performance to organizational outcomes (E-04)
Quality Gate	✓ All four Kirkpatrick levels have data sources identified and populated ✓ Level 4 metrics were defined before the program launched (not after) ✓ Tuning vs. replacing decision is data-driven, not intuition-driven ✓ Quarterly leadership report delivered on schedule ✓ Project knowledge base updated with Enhancement cycle data
AI Prompts **E-01 through E-04** Appendix A	**Primary Deliverables** Enhancement dashboard (weekly, monthly, quarterly cadence) Quarterly leadership report (Level 4 outcomes) Revised program content (tuning) or redesign brief (replacing) Updated project knowledge base for next cycle

Quick Reference: Where to Go in the Book

Foundation / RAG Setup: Chapter 4 (Prompt Engineering and Context Architecture) · Document Preparation Checklist

Phase IN in full: Chapter 6 · Prompts IN-01 through IN-11 (Appendix A) · Role chapters 10–13 for role-specific IN guidance

Phase D in full: Chapter 7 · Prompts D-01 through D-14 (Appendix A) · Chapters 11–12 for media and facilitation development

Phase I in full: Chapter 8 · Prompts I-01 through I-09 (Appendix A) · Chapter 12 for facilitation preparation

Phase E in full: Chapter 9 · Prompts E-01 through E-04 (Appendix A) · Chapter 14 for portfolio management at scale

Governance and ethics throughout: Chapter 17 · AI use policy · FACTS evaluation framework (Chapter 4) · SME review standard (Chapter 15)

INDIE

INvestigation-Design · Development Implementation · Enhancement

A Field Guide to AI-Augmented Learning and Development

www.ingramcontent.com/pod-product-compliance
Lightning Source LLC
LaVergne TN
LVHW052219200926
848481LV00001B/158

* 9 7 9 8 9 8 7 0 8 0 2 0 7 *